# Additive Manufacturing of Metals

**David J. Fisher**

Published by **Materials Research Forum LLC**
Millersville, PA 17551, USA

Published as part of the book series
**Materials Research Foundations**
Volume 67 (2020)
ISSN 2471-8890 (Print)
ISSN 2471-8904 (Online)

Print ISBN 978-1-64490-062-8
ePDF ISBN 978-1-64490-063-5

Distributed worldwide by

**Materials Research Forum LLC**
105 Springdale Lane
Millersville, PA 17551
USA
http://www.mrforum.com

Printed in the United States of America
10 9 8 7 6 5 4 3 2 1

# Table of Contents

## Introduction

Additive manufacturing, sometimes colloquially referred to as 3-dimensional printing or even as layered manufacturing, is a relatively new, flexible and increasingly important process for manufacturing or repairing structural components. Its use in 'rapid prototyping', using polymers as the building materials, is now fairly routine. Its use for building components out of metal alloys is of rapidly growing interest, but the very different properties of metals, as compared with those of polymers, mean that many problems remain to be solved. The 3-dimensional printing of thermoplastics is highly advanced and can easily create complex products, while the 3-dimensional printing of metals is still limited; the obvious snag being that the temperature-softening of thermoplastics into a readily formable state is absent in the case of metals.

It is therefore somewhat ironic that the apparently novel metal-forming technique pre-dates (figure 1) the polymer-based method unless of course one goes back even further and supposes that the patent in question was inspired by the ancient potters' technique of coiling long 'ropes' of clay so as to make vessels even without the benefit of a wheel. It is unclear from the patent whether the jug shown was watertight, as the patent title merely refers to a, "*Method of Making Decorative Articles*". On the other hand, the inventor also describes the products as being 'useful', so one has to read between the lines, as it were, of the somewhat ambivalent specification (figure 2). It is also unclear from the patent whether handles were to be formed as part of the process, or were to be added later. The diagram suggests the latter. A modern additive manufacturing process would certainly strive to create an object as a single piece.

The main driving force for the interest in additive manufacturing of metals comes from the aerospace, medical and dental industries and, consequently, the metals favoured for investigation have been titanium, nickel and iron alloys; particularly stainless-steels.

Metallic biomedical materials having a good corrosion resistance and mechanical properties are widely used in orthopaedic surgery and implanted dental devices, but they commonly cause so-called stress-shielding due to appreciable differences in elastic modulus between the implant and neighbouring bone. Given that the elastic modulus of a porous metal is lower than that of a solid metal however, it is possible to choose the pore characteristics so as to make the elastic modulus of a porous metal match more closely that of bone. An open porous metal with interconnected pores can moreover provide access to bone in-growth and thus strengthen the bond between bone tissue and an implant. Selective laser melting and electron-beam melting (to be described below) are important tools in additive manufacturing in that they lend themselves to the preparation of porous metal implants having a complex shape and fine structure. So-called topology optimization has been developed in order to enable the internal architecture of porous metals to be designed so as to exhibit specific mechanical properties.

*Figure 1. A jug fabricated from metal using layer-by-layer arc-welding according to a 1925 patent[1] assigned to Westinghouse*

Additive manufacturing of non-degradable materials such as titanium and CoCr alloys has been a great boon to clinical applications. The limitations of the investment casting of cobalt-based alloys are claimed to be less problematic following improvements in additive manufacturing, but the products are still likely to exhibit a mechanical anisotropy with respect to the build orientation which could impair their *in vivo* performance.

On the other hand, the similar treatment of biodegradable metals such as magnesium, iron and zinc is still under-developed. The evaporative loss of elements, and porosity, are common problems which arise in the additive manufacturing of zinc and magnesium due mainly to evaporation during melting by high-energy beams. Customized geometric shapes of surgical alloys having porous structures can here again be achieved accurately and efficiently by using laser powder-bed fusion.

*Figure 2. Detail of the construction method used in figure 1*

Even more risk-sensitive than the medical profession, the aerospace industry has been somewhat cautious towards applying additive manufacturing. This is because there still exist significant uncertainties concerning the quality of additively manufactured products. The laser-deposition of complex geometries, for example, can be difficult and therefore prone to error. For example, a study[2] of over 1000 nominally identical tensile tests explored the effect of process variability upon the mechanical property distributions of precipitation-hardened stainless steel produced using laser powder-bed fusion processes. This large dataset uncovered rare defects which affected only some 2% of the samples taken from a single source. These rare defects caused a marked decrease in ductility and were associated with an interconnected network of porosity. The safety concerns which are naturally associated with aerospace components therefore mean that additive manufacturing cannot be fully accepted for the creation of aircraft parts until it is deemed to be entirely reliable. The success of additive manufacturing thus depends not only upon its ability to compete with traditional manufacturing methods with regard to cost, but also with regard to the consistency of mechanical properties. The fact that additively manufactured metals can possess a high defect density means that account has to be taken of larger statistical variations as compared with those of conventional metals. In order to minimise time-consuming component-testing, computational methods are required which allow for topology and metal-type and which can precisely predict component failure, even for metals with statistically varying properties. Such predictions require automatic evaluation of large sets of material properties and their scatter.

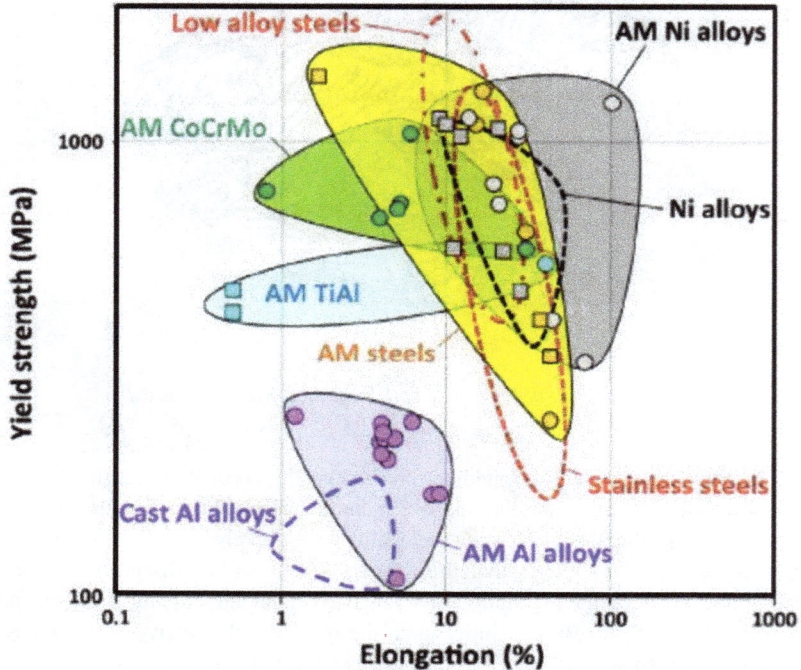

*Figure 3. Yield-strength versus elongation-to-fracture plot of a range of additively manufactured alloys. Note in particular the favourable offset of additively manufactured aluminium alloys as compared with those of conventionally cast aluminium alloys. Reproduced from: "Additive manufacturing of metals: a brief review of the characteristic microstructures and properties of steels, Ti-6Al-4V and high-entropy alloys", Gorsse, S., Hutchinson, C., Gouné, M., Banerjee, R., Science and Technology of Advanced Materials, 18[1] 2017, 584-610, under Creative Commons Licence.*

The additive manufacturing route nevertheless offers overwhelmingly tempting possibilities to this industry. Single-crystal nickel-based superalloy blades for instance are a key feature of the high-temperature gas turbines which are essential to aviation and power-generation because of their creep-resistance at high temperatures. Conventional manufacture of such blades is tedious and expensive. Additive manufacturing promises economic sustainability, automated manufacture and product uniformity. Figure 3 summarises the variety of materials which have been prepared using additive manufacturing.

In spite of the caution with regard to conventional aerospace use, plans are already being made for the use of additive manufacturing in free-fall as an adjunct to the logistical support of future space exploration. In order to exploit the technique in free-fall, appropriate equipment has already been designed.[3] The system comprises 8 laser-beams distributed in an annular array. These are to be focused on the substrate in order to create a molten pool while metal wire is fed vertically into the pool. It is planned to test the apparatus by using a so-called vomit-comet in free fall and thereby determine the effects of an apparent absence of gravity upon the molten pool behavior and the resultant microstructures and mechanical properties. Other apparatus has already been tested by using the drop-tower known as the 'Einstein Elevator'[4].

A theoretical investigation of free-fall manufacturing has been undertaken[5]. It was pointed out that, in space, the additive manufacturing environment will experience almost zero gravity, as well as perhaps near-vacuum conditions. Computational fluid dynamic transient models were therefore developed for the effect of extreme low-gravity and low-pressure effects upon laser metal deposition process, with wire feeding. Experimental validation of the model was attempted by performing deposition at various orientations on Earth. The results showed that surface tension would dominate melt-pool dynamics under reduced gravity and that consequent irregular deposition tracks would appear. A reduction in wire volume deposited per unit length, as compared with that on Earth, was predicted to minimise that effect. It was also predicted that material vaporization was more likely to occur as the pressure was reduced, but that this could be avoided by decreasing the laser power or increasing the scanning speed. Such is the enthusiasm for the use of additive manufacturing in space that attention is already being paid to the possibility of its use on other planetary bodies, and experimental studies have already been made[6] of the mechanical behaviour of simulated lunar regolith material and its treatment using laser-based powder-bed fusion methods.

The methods which are used for the additive manufacturing of metals follow the basic concept of the patent shown: that is, layer after layer is added sequentially in order to generate the desired 3-dimensional shape. Obvious advantages of this method, as compared with the traditional, and usually 'subtractive', manufacturing methods are that the process is fast and there is no need for expensive tooling of moulds or dies. The old patent did not attract much interest, probably because it describes a clearly labour-intensive method. But thanks to modern technology, the process is now capable of fabricating complex parts whose intricacies have been stored as digital data, as in the case of computer-aided design, and where the practical work can even be performed by robots. It is interesting to note here that Ralph Baker, author of the 1925 patent, had already foreseen the possibilities. As he put it, "*I need not necessarily manipulate the electrode by hand. If the electrode is to be manipulated to form a number of articles having like contour, pantographic or other apparatus, such as is used by engravers, may be employed*".

The main competitor is that of traditional casting: both techniques being capable of producing parts which contain internal holes, and with the surface finishes being generally similar. A principal concern in industrial applications is the range of materials that can be processed while ensuring that the resultant surface roughness and mechanical properties are as least as good as those of the equivalent cast or wrought materials.

Unevenness of the surface shaped by overlapping beads has sometimes been observed during metal additive manufacture. Any uneven surface can have a cumulative effect upon the accuracy of the component in the vertical direction: what one might call a 'princess and the pea' situation. Milling of the face of each layer has been essayed in order to solve the problem, but this results in a waste of material and time. Rolling has also been suggested as an alternative solution, but it is generally better to ensure that such unevenness does not arise in the first place.

Another advantage is that the traditional production of very complex parts requires them to be broken down into smaller parts, thus costing further material and time, and yet more processing may be required when the parts have been completed and assembled. Additive manufacturing can produce components having the required shape, and surface properties, in a single step.

A number of methods have been developed by which to add the sequential layers. One of the most favoured technologies in the aerospace industry is laser metal deposition. This involves the use of a moving laser to melt metal powders which are coaxially ejected near to the laser focal-point. This creates a molten pool on the substrate which traps the powders. The solidified tracks then serve to construct complex components layer-by-layer. The resultant properties of the deposited material depend upon factors such as the shape of the powder particles, the supply-rate of the powder and the laser-parameters. Considered in more detail, the process depends for success upon the powder flow, the laser-powder interaction, the formation of a suitable molten pool by the laser irradiation and proper solidification of the melt pool. The adhesion properties of powder particles can greatly affect the quality of parts made by additive manufacturing processes.

Accurate experimental characterization of the adhesion of a single microparticle has however been very difficult, due mainly to the problems associated with the controlled micro-scale handling of such particles.

The processing parameters of laser additive manufacturing have a critical effect upon the solidification microstructure, particularly with respect to the grain size. A three-dimensional model which took account of heat-transfer, phase changes and Marangoni convection showed that the cooling rate increased and the grain-size decreased, from 8.7 to 4.7µm, upon increasing the scanning speed from 2 to 10mm/s. The cooling-rate decreased and the grain-size increased with increasing laser power and powder feed-rate.

Additive manufacturing naturally generates less waste than does conventional manufacturing. In the case of selective laser melting, it might even be feasible to re-cycle side-streams back into the feedstock. Powder might be prepared from 100% scrap feedstock either by mechanical milling of agglomerated powder residue or by gas atomization of solid scrap; even without using extra alloying to compensate for element-loss. Preliminary tests have suggested that recycled powders can possess properties which are within specifications.

Most of the current research focuses on the optimization of the processing parameters of pre-alloyed powder feedstocks of established alloys used for structural applications: that is, mainly stainless steels, titanium-based alloys and nickel-based alloys. The aim is to obtain properties which exceed those of the conventionally processed materials. A further advantage of additive manufacturing however, is that the properties can be caused to vary locally. It might be used, for example, to deposit magnetic alloys.

Laser additive manufacturing can be used to produce components which are microstructurally and compositionally gradated. Blocks consisting of a sequence of roughly 500µm-thick tool-steel layers, each having a different (iron-based and nickel-based) chemical composition, can be deposited layer-wise on a substrate. The layers between them are made to consist of blends of the two materials, with the volume fractions varying (80:20, 60:40, 40:60, 20:80) from layer to layer. The bulk alloy is then hot-rolled and heat treated. The aim is to achieve the rapid exploration of novel compositional alloy blends. It can be used to identify new alloys possessing any desired combinations of tensile strength and ductility.

Selective laser sintering can potentially solve problems related to the production of tungsten carbide particles embedded in a tough metallic binder, as the method can resolve difficulties arising from differences, such as thermal conductivity, in the physical properties of the carbide and the metallic binder. The processing of hard refractory ceramic coatings in a metal matrix is difficult due to factors such as delamination or cracking due to large property differences at the interphase, and non-uniform distributions due to poor mixing. These difficulties can be avoided if the coating exhibits a gradual change in properties from the surface to the interior. Additive manufacturing is capable of furnishing the required compositional control.

As well as compositional variations, the use of built-in anisotropy is an interesting design choice when creating products, especially in the fields of aerospace and biomedicine where components may be exposed to anisotropic stress-fields. Products should then be anisotropic along the functional axis. Powder-based metal additive manufacturing permits the control of a wide range of anisotropies, from crystallographic texture to pore structure.

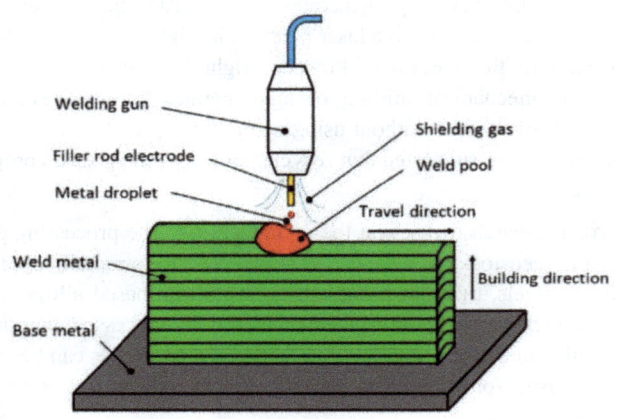

*Figure 4. Schematic diagram of the arrangement in the wire arc additive manufacturing process. Note the essential similarity to the situation depicted in figure 2. Reproduced from "Fabricating superior NiAl bronze components through wire arc additive manufacturing", Ding, D., Pan, X., van Duin, S., Li, H., Shen, C., Materials, 9, 2015, 652, under Creative Commons Licence.*

Powder is generally used as a filler material for the laser additive manufacturing of metallic components, but it is unsatisfactory with regard to powder-milling costs, deposition efficiency, material-utilization ratio and environmental adaptability. The packing density of metal powders is an important factor as it directly affects the physical and mechanical properties of the finished products. In order to achieve the most efficient packing of a powder, it is well known that differing grades of the powder must be mixed together in proportions which minimize the presence of voids. It has been shown that packing the coarser grains first gives higher-density powders and decreases the incidence of balling (coalescence) defects in the finished product. A simple model exists which predicts those volumetric fractions of differing powder grades that yield the highest powder density.

Wire-feed additive manufacturing has attracted increasing attention, but poor melt-stability of the filler wires impedes its development. There are also problems of poor material flexibility, poor dimensional accuracy and limited energy-efficiency. Pre-heating of the wires has alleviated some of the drawbacks but it can still be difficult to obtain adequate properties without applying further bulk heat-treatment. The structures and properties of components made by wire arc additive manufacturing depend upon process parameters such as the arc power, travel speed, wire diameter and wire feed-rate. Exploring this parameter-space, in order to identify that selection which will produce defect-free structurally-sound components can be expensive and time-consuming.

The wire arc process exploits well-established arc-welding technology to melt wire, by arc discharge, which thus locally adds material to the molten pool. It is also perhaps the closest in spirit to the 1925 patent (figure 4). By changing the material during the process, more than one type can be used simultaneously in the manufactured component. For example, stainless steel and a nickel-based alloy can be combined such that the mechanical properties of the manufactured alloy are comparable to those of bulk material. It could be possible even to construct components in which the surfaces and inner structures were made from different materials.

Arc-based additive manufacturing processes offer the additional advantages of almost unlimited assembly space, higher deposition rates and a better utilization factor of raw materials. The disadvantages include the limited range of available wire types and the fact that the wire feed-rate is directly linked to the heat input. It is also not possible to create multi-material structures *in situ* using a single heat-source. Advantages are offered by the 3-dimensional plasma-metal deposition method, which is based upon a plasma powder-deposition process. Structures can be fabricated by using welding-robots, with the path-control governed by computer-aided design files. This results in an increased flexibility with regard to material selection, and the building of gradated structures.

So-called metal big area additive manufacturing is a wire arc method which uses a feedback correction-based approach, here again exploiting computer technology, to minimize the dynamic nature of the welding process and allows for the increasing height of the deposits. This process can be used to create simple-geometry thin-walled specimens by, for example, using C-Mn steel welding wire. Tensile tests have revealed isotropy of the tensile and yield properties with respect to the build direction, but there can be a large scatter in Charpy impact-test results. The microstructure comprises mainly homogeneous ferrite grains, plus some pearlite, with changes in the morphology and grain size at the interface between the built object and the base plate. This process generally imparts stable isotropic weld-like mechanical properties to the deposit, with a precise shape being guaranteed by the feedback software.

Another popular route is powder-bed additive manufacturing, in which the feedstock is a powder that forms an homogeneous bed having the cross-section of the manufactured object and which is then fused. A new powder layer is then distributed and the process is repeated. This is perhaps the metal-based process which is closest in principle to the now-familiar polymer-based 3-dimensional printer. This approach can be incorporated into the selective laser sintering/melting process, the electron-beam melting process and the binder jetting process.

Selective electron-beam melting, a powder-bed metal additive technique, has been used to create nickel-based monocrystalline superalloys. The high vacuum environment and almost unidirectional thermal gradient which are naturally associated with this process make it the most promising additive manufacturing technique for monocrystalline superalloy production. Microstructural examination reveals that predominant columnar grains, aligned in the build direction, formed a strongly textured solid, although some cracking occurred along the grain boundaries. The layouts of some of the most popular additive manufacturing methods are compared in figure 5.

Among additive manufacturing processes, the binder jetting method has the advantage that it allows a wide choice of material selection and design. On the other hand, the arrangement which determines the direction in which the binder is injected during layer-stacking can affect the final properties. In this process, metal parts are made by first jetting a binder into a powder-bed, and then sintering the resultant green part while the binder is removed and the metal particles are fused. The properties of the sintered part can be increased when nanoparticles are suspended in a solvent-based organic binder because the ink-jetted nanoparticles can reduce sintering shrinkage and increase the strength. It is also possible to use a nanoparticle suspension without organic adhesives as a means of binding metal powder-bed particles together. After being deposited into the powder-particle interstices, the jetted nanoparticles are sintered at a temperature which is lower than the powder sintering temperature so as to impart strength to the printed green product. Unlike organic binders, the use of jetted nanoparticles provided a permanent bonding which improved the structural integrity of the parts during sintering.

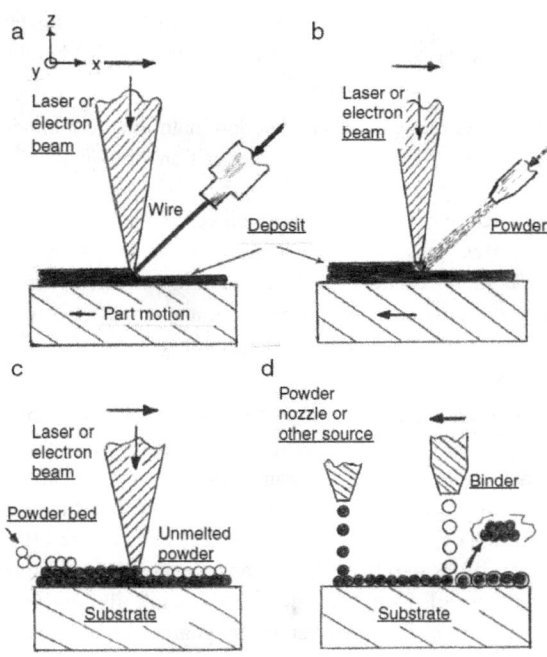

*Figure 5. Comparisons of additive manufacturing layouts: a. laser or electron-beam melting with wire feed, b. laser or electron-beam sintering with possibility of mixed feeds, c. powder-bed with electron-beam or laser selective melting, d. binder jet powder process requiring subsequent sintering to bind metal powder. Reproduced from, "3D metal droplet printing development and advanced materials additive manufacturing", Murr, L.E., Johnson, W.L., Journal of Materials Research and Technology, 6[1] 2017, 77-89, under Creative Commons Licence.*

A significant bottleneck is associated with powder spreading, as mechanical arching badly affects both product quality and production speed. There can be transient jamming of gas-atomised metal powders during spreading. The particles are very friction-prone as they have asperities and tend to jam in narrow gaps. The characteristic size, D90, for which 90% of particles by number are smaller than this value, is used as the particle dimension which accounts for jamming. The latter phenomenon is observed in the form of empty patches over the work surface. Its frequency and period can be characterised as a function of the spreader gap-height, and expressed as a multiple of D90. The probability of formation of empty patches and their mean length, where the latter indicates the duration of jamming, increases markedly with a decrease in gap height. The collapse of mechanical arches leads to particle bursts.

Beam- and wire arc-based methods clearly predominate for the moment, although many others are being developed. Layer-by-layer deposition obviously offers incomparable advantages to critically-important industries, such as aerospace. Companies in that field use large quantities of expensive high-strength nickel and titanium alloys, and traditional manufacturing processes generate considerable amounts of waste. Shaped metal deposition, one of a range of additive manufacturing methods, creates near-finished components by means of tungsten inert-gas welding and is particularly useful for titanium alloys because the latter are difficult to shape using the conventional methods which produce extensive amounts of expensive swarf.

High-speed metal particle cold-state impact-based additive manufacturing has been heralded as solving the problems which afflict the metal-part free-forming process, such as high energy consumption and high thermal residual stresses. In this process, high-pressure (1.5 to 3.5MPa) nitrogen is used to accelerate metal particles so as to impact the substrate and thus form a deposit at room temperature. Experimental results show that the diameter and thickness of the deposited splats can be controlled by the pressure, while the width and layer-thickness of deposited lines can be controlled by the pressure and the nozzle velocity. Effective cohesion between layers guarantees the ability of the process to produce objects layer-by-layer.

As noted earlier, the 3-dimensional printing of thermoplastics is well-advanced and can easily create complex geometries while the 3-dimensional printing of metals remains fairly limited. The difference can be traced to the fact that thermoplastics continuously soften with temperature into a readily formable state whereas normal metals do not. Unlike conventional metals however, bulk metallic glasses exhibit a supercooled liquid region and continuous softening upon heating, in an analogous fashion to that of thermoplastics. These glasses are also suitable for extrusion-based 3-dimensional printing via fused filament fabrication. By exploiting the supercooled-liquid behaviour of metallic glass, 3-dimensional printing can be achieved under similar conditions to those applicable to thermoplastics. Fully-dense amorphous metallic-glass parts can be 3-dimensionally printed under ambient environmental conditions and produce high-strength metal.

In the additive friction stir process, wrought metal is deposited onto a metallic substrate and involves both solid and powder filler being fed in layer-by-layer. Shear-induced interfacial heating and severe plastic deformation are then used to produce metallurgical bonding between the powder particles and layers. This process can be more economical and energy-efficient when making near-net shape components from thermally sensitive metallic materials.

The semi-solid metal forming method, combined with deposition processes, is familiar as being the most common additive manufacturing process which is applied to polymers. Semi-solid metal forming is a promising near-net shape technology which offers several advantages: such as, porosity-free products, reduced shrinkage, controlled microstructure and excellent mechanical performance. On the other hand, the complicated time-dependent behavior of semi-solid metal forming makes it difficult to use. A so-called strain-induced melt-activated process has been applied to a low melting-point Sn-Pb alloy in order to obtain the required globular feedstock microstructure. Semi-melted alloy is then deposited onto a moving substrate so as to build a metallic part layer by layer. Acceptable metallurgical layer-bonding is obtained at the interface of the deposited layers, imparting good mechanical properties to the parts. The semi-solid metal extrusion and deposition process can be considered to be a non-powder metal additive manufacturing method. Unlike other methods, it involves the extrusion of a metallic filament, in a semi-solid state, so as to build a product layer-wise under computer control: a thixo-extruder, mounted on a 3-axis table, can deposit semi-solid slurry layer-by-layer. Correct design of the thixo-extruder is critical for success, together with the magnitude of the semi-solid fraction. Alloys with solid fractions ranging from 0.3 to 0.6 can be thixo-extruded.

Few additive manufacturing processes have been developed for creating metal-matrix composites with a ceramic reinforcement. These processes usually employ various mixing techniques to combine metal and ceramic powder particles for finishing by, for example, selective laser melting. A modified additive manufacturing method for the preparation of metal-matrix composites is the thermal decomposition of salts. In this method, inorganic salts are printed onto a metal powder-bed in order to make a green part. The latter then undergoes bulk sintering. During the sintering, the printed inorganic salts decompose into fine ceramic particles to form the metal-matrix composite. This process is capable of generating composite structures containing uniformly distributed and dispersed ultra-fine ceramic particles in a metal matrix; with fewer limitations and lower cost as compared with those of other additive manufacturing techniques.

Fused filament fabrication is a well-known tool-free 3-dimensional printing technology that can produce complex components having free-form surfaces. Based upon the extrusion of a thermoplastic filament through a small nozzle, the associated machines are characterized by having relatively little complexity. By using a filament having a high metal powder loading, it is possible to print green parts that can be debinded and sintered.

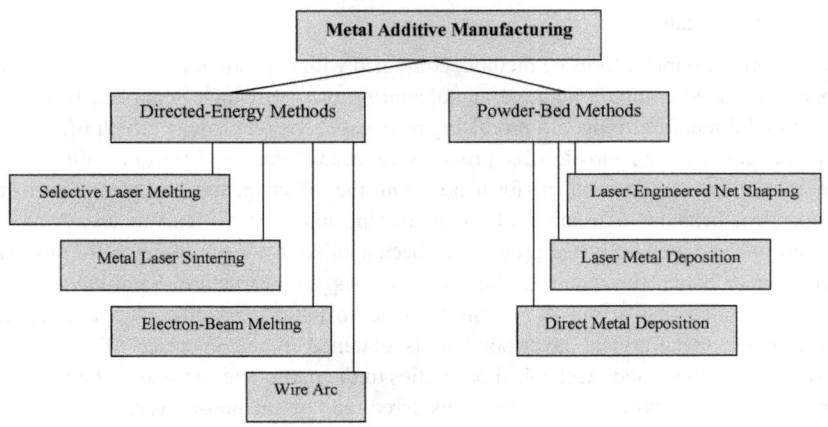

*Figure 6. Summary of the currently predominant additive manufacturing processes*

Metal foils can be bonded together by using a very high ultrasonic energy and metal atoms, heated by the high-frequency vibrations diffuse from one surface to another to form a solid-state metallurgical bond. When compared with other additive manufacturing techniques, such high-power ultrasonic bonding has many advantages: such as, low temperatures, little deformation, high consolidation-rate and environmental friendliness. Ultrasonic additive manufacturing is a 3-dimensional printing process which uses sound to merge layers of metal drawn from featureless foil stock. The process produces true metallurgical bonds with full density and is applicable to many different metals.

Micro-coating is a novel technology which can build near-net components layer-by-layer, using a crucible and a nozzle instead of a weld-head and wire-feeder to supply material for shaped metal deposition. A pneumatic system adjusts the liquid-metal flow-rate, and the layer height is controlled by the distance between the nozzle and the substrate.

A wide range of metals can be dissolved, and re-deposited from liquid solvents, by electrochemical means. This principle can be exploited for additive manufacturing, via the focused electrohydrodynamic ejection of metal ions dissolved from sacrificial anodes, and their reduction to elemental metal form on the substrate. This so-called electrohydrodynamic redox printing technique permits the direct ink-free construction of polycrystalline multi-metal 3-dimensional structures without needing subsequent processing. *Ad hoc* switching of two metals issuing from a single multichannel nozzle permits chemical features, with a size of less than 400nm and a spatial resolution of 250nm, to be printed at speeds of up to 10voxels/s.

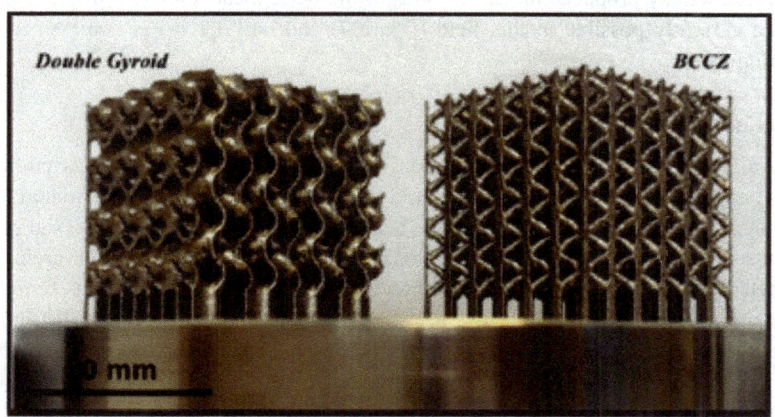

*Figure 7. An example of what is currently possible in the field of additive manufacturing: smooth, finely-detailed constructs prepared using a commercial printing device. Reproduced from, Nesma T. Aboulkhair, et al., Progress in Materials Science, https://doi.org/10.1016/j.pmatsci.2019.100578 "3D printing of Aluminium alloys: Additive Manufacturing of Aluminium alloys using selective laser melting", Aboulkhair, N.T., Simonelli, M., Parry, L., Ashcroft, I., Tuck, C., Hague, R., Progress in Materials Science, 2019, 100578 under Creative Commons Licence.*

The possibility of the electrochemical μ-additive manufacture of nanotwinned metals has been mooted, in which the presence of parallel arrays of twin boundaries would improve the mechanical and electrical properties. The microstructures of metals which are printed by using the microscale localized pulsed electrodeposition process can be controlled *in situ* during 3-dimensional-printing. In particular, electrochemical process parameters can be used to control the density and orientation of the twin boundaries, as well as the grain size. Micro-compression testing of 3-dimensionally-printed micro-pillars shows that such control over the microstructure directly affects the mechanical properties of the printed metal.

Leaving aside these last few relatively-specialized techniques, the already slightly bewildering range of mainstream processes is summarized in figure 6. The main purpose of the present work however, is not to give a detailed description of the various processing methods, but rather to summarise the various materials which are amenable to additive manufacturing, and how their properties are improved, or not, when compared with conventionally produced metals. It is inspiring to see meanwhile just what *tours-de-force* are currently possible in this field (figure 7) and how far it has come since that pioneering patent.

**Aluminium**

A solid-state friction stir technique of additive manufacturing, based upon friction stir welding, was used[7] to build multi-layered stacks of aluminium-based material. A transition zone formed near to the interface between two layers, and fine equiaxed grains were observed due to overall dynamic recrystallization. A difference in grain size nevertheless existed throughout the build direction and a band of coarse grains formed in the transition zone. A similar change in precipitate morphology, size and distribution occurred from the top to the bottom. The microhardness thus varied greatly, with a maximum of 115HV at the top. The tensile strengths of all of the slices increased and the elongation decreased slightly as compared with the aluminium substrate. The top possessed the best mechanical properties, due probably to the fine grains and the nature of the precipitates.

The acoustoplastic metal direct-write method can be used to prepare mm-scale 3-dimensional aluminium articles. Induced inter-layer and intra-layer mass-transport results[8] in metallurgical bonding. During processing, there was a temperature increase of 5C above the nominal process temperature of 25C. Acoustic-energy induced microstructural changes could be seen in the product. This non-melt room-temperature 3-dimensional metal printing approach was capable of producing materials having a more than 99% density. High-power ultrasonic additive manufacturing was applied[9] to the preparation of aluminium-copper composites. Multi-laminated ultrasonic-consolidated samples were prepared at various temperatures, using T2 copper foil and AA1060 aluminium foil. Peel tests of Al/Cu and Cu/Al interfaces showed that the Cu/Al interface had a higher peel strength. The mechanical strength of the Al/Cu interface increased with increasing preparation temperature, and the Al/Cu interface could form an effective solid-phase bond under ultrasonic solicitation at a preparation temperature of 100C. The width of the interdiffusion layer at the interfaces between the metals was up to 6μm. The highest mechanical strength was 26.23N/mm, and the fracture surfaces had the form of a torn prism with quasi-cleavage failure.

The cold spray additive manufacturing method was used[10,11] to produce various aluminium-matrix composites. A hybrid, hot-compression plus hot-rolling, post-deposition treatment was used to improve the mechanical properties of cold spray additively manufactured Al-B$_4$C composites. As-deposited samples were first subjected to a 30% thickness reduction by hot compression at about 500C, followed by hot rolling to 40% thickness reduction in 2 passes. Following a hybrid post-deposition treatment involving a 70% accumulated thickness-reduction, the aluminium grains of the matrix were greatly refined due to continuous dynamic recrystallization and simultaneous geometric dynamic recrystallization. The number of interfacial defects was markedly reduced, and the nature of the Al|Al and Al|B$_4$C interfacial bonding changed from mechanical interlocking to metallurgical bonding; thus improving the transfer of an applied load to the uniformly dispersed bimodal B$_4$C particles. The ultimate tensile strength and elongation of as-deposited samples consequently increased simultaneously from 37 to 185MPa and from 0.3 to 6.2%, respectively. The cold spray technique was also used[12] to deposit 5mm-thick SiC-reinforced aluminium-matrix composites. Specimens which were in the as-sprayed condition fractured in a brittle manner and had a tensile strength of 85MPa which resulted mainly from intensive work-hardening. Following heat treatment at 200, 300, 400 or 500C, the tensile strength of 104MPa and plasticity of 1.5 to 5.2% were improved due to coarsening of the pure aluminium via recrystallization, recovery and grain-growth. The main fracture mode in the heat-treated state was ductile.

*Table 1. Tensile properties of selective laser melt prepared Al-3Ce-7Cu alloy*

| Condition | Temperature (C) | Direction | YS (MPa) | UTS (MPa) | El (%) |
|---|---|---|---|---|---|
| as-prepared | 25 | XY | 274 | 459 | 4.4 |
| annealed (200C, 3h) | 25 | XY | 247 | 386 | 3.8 |
| annealed (300C, 1h) | 25 | XY | 231 | 383 | 4.2 |
| as-prepared | 25 | Z | 257 | 349 | 1.1 |
| annealed (200C, 3h) | 25 | Z | 228 | 346 | 1.4 |
| annealed (300C, 1h) | 25 | Z | 212 | 297 | 1.0 |
| as-prepared | 250 | XY | 176 | 186 | 17.6 |
| annealed (300C, 1h) | 250 | XY | 174 | 192 | 18.8 |
| as-prepared | 250 | Z | 160 | 184 | 11.2 |
| annealed (300C, 1h) | 250 | Z | 169 | 200 | 12.2 |

The friction stir additive manufacturing process has been used[13] to create various aluminium-based composites, including Al-10vol%SiC, Al-20vol%SiC, Al-30vol%SiC, AA6061-6.1vol%W and AA6061-22vol%Mo[14]. Fully-dense parts could be obtained which were free from porosity, wormholes and cracks. The reinforcing particles in as-deposited samples had an essentially uniform distribution, with improved mechanical properties: the hardness of AA6061 was increased from 49.7HV to 357.2HV by the addition of 22vol%Mo.

*Al-Ce*

Selective laser melting was used[15] to produce samples of the heat-resistant alloy, Al-3Ce-7Cu, with fine $Al_{11}Ce_3$ and $Al_{6.5}CeCu_{6.5}$ eutectic phases being found in the microstructure. Annealing at 250 to 400C led to a decreased hardness, but the latter was higher after annealing at 350 or 400C than after annealing at 250C (table 1). This was due to the precipitation of nanosized particles. The low hardness observed following quenching and aging at 190C was caused by quench stress-relief and by the absence of age-hardening, due to poor solid-solution. The as-prepared yield stress, ultimate tensile strength and elongation were 274MPa, 456MPa and 4.4%, respectively.

*Al-Cu*

When cold metal transfer was used[16] to make a thin-wall part from 205A (Al-5wt%Cu) alloy, an increased ultimate tensile strength of the as-deposited material, as compared with that of the as-cast equivalent, was attributed mainly to the presence of refined α-aluminium grains rather than that of the dot-like $Al_2Cu$ phase. A slightly inferior plasticity was found, and this was attributed mainly to micro-scale spherical pores and a relatively high (3.51%) porosity.

In the wire arc additive manufacturing of AA2319 components, Al-6.3%Cu deposits were produced[17] by using cold metal transfer variants: pulsed cold metal transfer and advanced cold metal transfer. Thin walls and blocks were used for the deposition. As compared with pulsed cold metal transfer in the thin-wall mode, advanced cold metal transfer and blocks could reduce the tendency to porosity in the aluminium alloy. The microstructure depended upon the deposition mode and the cold metal transfer variant. The microhardness of thin-wall samples was about 75HV from the bottom to the middle, and gradually decreased towards the top. The microhardness ranged from 72 to 77HV and varied periodically, in the block mode, determined by the microstructure. In further work, aluminium-copper (ER2319) and aluminium-magnesium (ER5087) wires were used[18] as filler metals to build Al-Cu-Mg components. Differing compositions could be obtained by adjusting the wire feed-rates. The microstructures of the Al-Cu-Mg deposits comprised mainly coarse columnar grains and fine equiaxed grains having a non-uniform distribution. With increasing copper content but decreasing magnesium content, the strengthening phase changed to $Al_2Cu$ plus $Al_2CuMg$, from $Al_2CuMg$, and the microhardness increased. The ultimate tensile strength was about 280MPa in both the horizontal and vertical directions. The yield stress increased from 156 to 187MPa, while the elongation decreased from 8.2 to 6%. The fractures were brittle.

The reactive additive manufacturing method has been used[19] to produce aluminium-matrix composites. In the case of AA2024 composites, the latter possessed some 2.3 times the yield stress, 1.7 times the ultimate tensile strength, 1.4 times the elastic modulus and 3.8 times the wear-resistance of AlSi10Mg. These properties were also largely retained at high temperatures, with a less than 7% reduction in yield stress at 150C.

Double-wire arc additive manufacturing, using variable-polarity gas tungsten arc welding methods and Al-6.3Cu plus Al-5Mg wires, has been applied[20] to the fabrication of Al-Cu-Mg products. Excellent wall samples could be produced, and the deposition efficiency was certainly improved by using the double-wire process. Aluminium alloys containing various amounts of copper and magnesium could be obtained by adjusting the feed-speeds of the two wires. The resultant deposit microstructures were composed mainly of columnar and equiaxed dendrite grains having a non-uniform distribution. The microhardness of the as-deposited Al-Cu-Mg alloy was 900 to 1000MPa, while the ultimate tensile strength, yield stress and elongation of the as-deposited alloy were 286MPa, 183MPa and 4.5%, respectively.

Wire arc additive manufacturing methods were used[21] to form Al-Cu-Mg alloys by feeding-in aluminium-copper and aluminium-magnesium wires in tandem. Macro- and micro-cracks were identified as being intergranular solidification cracks. A survey of the cracking susceptibility as a function of the copper and magnesium contents revealed that compositions with 4.2 to 6.3% of copper and 0.8 to 1.5% of magnesium were less susceptible to cracking during solidification which ended in isothermal ternary eutectic reaction. A higher microhardness tended to reduce the cracking susceptibility. Another survey indicated that a higher wire-feed speed caused a higher heat input but a lower density of the deposited alloy; markedly increasing solidification cracking. The maximum susceptibility occurred when the microhardness was below 95HV and the heat input was greater than 200J/mm. Microcracks could initiate from the interlayer equiaxed-grain zone if insufficient liquid feeding prevailed during deposition.

*Al-Mg*

Wire arc additive manufacturing has been used[22] to fabricate aluminium-magnesium thin-wall components by feeding-in Al-5%Mg wire and adding titanium powder. The presence of the $Al_3Ti$ which was thereby formed provided heterogeneous nucleation cores, which then promoted the refinement of interlayer grains, a transition from columnar to equiaxed and an increase in the fraction of low-angle grain boundaries. Upon comparing thin-wall components, with and without titanium powder, the ultimate tensile strength and elongation of those with titanium were found to have increased in the vertical and horizontal directions. The microhardness of the interlayer was increased by 5 to 10HV.

Fully dense components made from Al-12Si and Al-10Si-1Mg have been produced[23] by using a continuous-wave selective laser melting method. Pulse selective laser melting was expected to permit better control of the heat input. In the case of Al-12Si, silicon refinement to below 200nm was possible, with a density of up to 95% and a hardness of more than 135HV.

*Table 2. Texture components (%) of wire arc prepared and rolled AA5087*

| Condition | Texture Type | Component (%) |
|-----------|--------------|---------------|
| as-prepared | brass | 8.08 |
| as-prepared | S | 5.92 |
| as-prepared | copper | 8.58 |
| as-prepared | Goss | 7.38 |
| as-prepared | cube | 3.65 |
| as-prepared | random | 66.39 |
| as-prepared | total rolling | 22.58 |
| rolled, 15kN | brass | 6.09 |
| rolled, 15kN | S | 6.42 |
| rolled, 15kN | copper | 8.44 |
| rolled, 15kN | Goss | 9.12 |
| rolled, 15kN | cube | 4.19 |
| rolled, 15kN | random | 65.74 |
| rolled, 15kN | total rolling | 20.95 |
| rolled, 30kN | brass | 11.19 |
| rolled, 30kN | S | 10.4 |

| | | |
|---|---|---|
| rolled, 30kN | copper | 13.7 |
| rolled, 30kN | Goss | 6.4 |
| rolled, 30kN | cube | 4.17 |
| rolled, 30kN | random | 54.14 |
| rolled, 30kN | total rolling | 35.29 |
| rolled, 45kN | brass | 25.51 |
| rolled, 45kN | S | 21.7 |
| rolled, 45kN | copper | 13.12 |
| rolled, 45kN | Goss | 14.63 |
| rolled, 45kN | cube | 1.84 |
| rolled, 45kN | random | 23.2 |
| rolled, 45kN | total rolling | 60.33 |

The double-pulsed arc additive manufacturing process was used[24] to produce high-strength Al-Mg components in 0, 1, 3, 5, 7 or 9Hz multilayer deposition experiments. Porosity was clearly eliminated, as compared with samples prepared by using the cold metal transfer process. The appearance of the bead of deposited aluminium varied with low frequency and, at a frequency of 3Hz, its appearance was optimum. The average grain size varied markedly with frequency and, at 1Hz, was at its minimum. The deposited material exhibited its greatest microhardness and highest ultimate tensile strength at a frequency of 3Hz. Another sort of frequency can affect wire arc additively manufactured products. That is, imperfections in aluminium can be suppressed by workpiece-vibration[25]. The vibration breaks off dendrite arms by imposing bending stresses upon them, and the grain size is refined by the formation of additional nuclei. The vibration could reduce the average grain size by a maximum of 22.5% as compared with non-vibrated samples. The workpiece vibration also induced strong stirring effects in the molten pool, which thus removed the fine-grain zone at the interlayers and suppressed porosity. Finally, optimum treatment increased the ultimate tensile strength of vibrated samples up to 343MPa; exceeding that (<340MPa) of wrought material.

Interlayer rolling, using loads of 15, 30 and 45kN, was applied[26] to AA5087 (Al-Mg4.5-Mn) alloy samples which had been prepared by means of wire arc additive manufacturing. As compared with the as-deposited material, the average microhardness, yield strength (figure 8) and ultimate tensile strength of 45kN-rolled alloys attained 107.2HV, 240MPa and 344MPa; enhancements of 40, 69 and 18.2%, respectively. Initially coarse grain-structures were greatly refined, and had a clear rolling texture following deformation (table 2). The improvements arose mainly from deformation-strengthening, grain-refinement and solution-strengthening. The elongation of the rolled samples remained above 20%, and the plasticity was not obviously impaired as compared with as-deposited alloy. This was twice that of commercial wrought Al-Mg alloy of similar composition. The excellent plasticity was attributed to grain-refinement, pore-closure, pore reduction and grain recrystallization during wire arc re-heating.

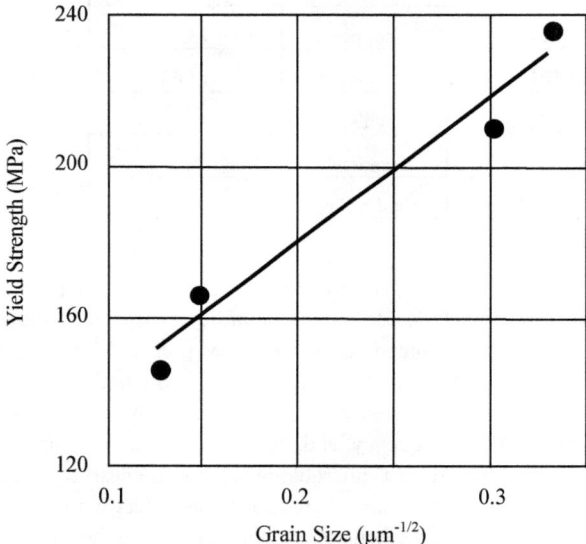

*Figure 8. Hall-Petch relationship for the yield strength
of rolled wire arc manufactured AA5087*

A study was made[27] of the use of wire arc additive manufacturing to deposit AA5183 aluminium alloy onto 20mm-thick AA6082-T6 plate by using conventional gas metal arc welding. Hardness values of about 75kg/mm$^2$ were found in the horizontal plane, and of between 70 and 75kg/mm$^2$ in the vertical plane; down to the AA6082 with a value of 100kg/mm$^2$. The yield and tensile strengths were 145 and 293MPa, respectively, with the lowest values found for the through-thickness direction. The ductility was high for orientations parallel to, and perpendicular to, the layer deposition direction. The effect of the arc-mode upon the properties of AA5183 prepared by cold metal transfer was investigated[28]. The highest tensile strength was found for samples built using the CMT+A process, which also had the smallest average pore size. The average tensile strengths of CMT+A built samples were 296.9 and 291.8MPa in the horizontal and vertical directions, respectively. The differences in tensile strength in the horizontal and vertical directions were due mainly to pores existing at the interfaces between each deposited layer. Walls of ER5183 (Al-[4.3 to 5]Mg-[0.5 to 1]%Mn) were wire arc additively manufactured[29] as various layers using variable polarity cold metal transfer. In the top layer, the grain size increased from 37 to 65μm, before stabilizing at about 41μm. This corresponded to a decrease in microhardness from 98.7 to 76.4HV. As the number of deposited layers increased, anisotropy in the bottom layers lessened and low-angle grain boundaries could increase the microhardness to 96.1HV before stabilizing at about 81HV due to heat accumulation.

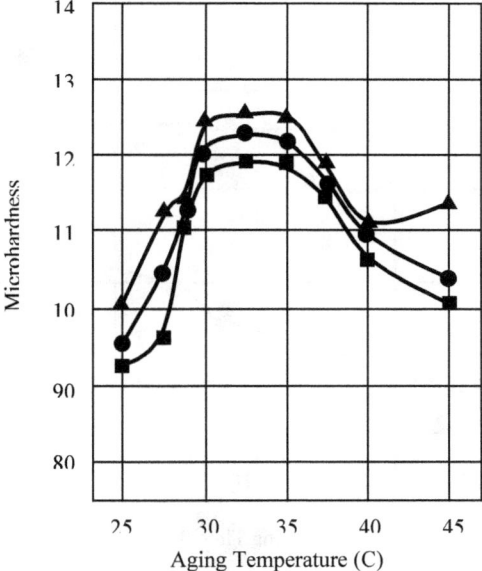

*Figure 9. Microhardness of selective laser melt (scan speed: 800mm/s) Al-3.02Mg-0.2Sc-0.1Zr as a function of aging temperature (4h) for laser powers of 400W (triangles), 300W (circles) or 200W (squares)*

A study of the wire arc additive manufacture of 5A06 (Al-6Mg-Mn-Si) using 1.2mm diameter filler wire showed[30] that the minimum angle, and radius-of-curvature, which could be made by using the wire arc method were 20° and 10mm when the layer-width was 7.2mm. There was some isotropy of properties in the build direction and the perpendicular one. When loading in directions parallel to, and perpendicular to, the texture orientation the tensile properties were anisotropic; with a difference of 22MPa. The layer-height fell rapidly, from a first layer of 3.4mm, and remained at 1.7mm after the 8th layer. Fine dendritic and equiaxed grains were found inside the layers, with the coarsest columnar dendrite structure being at the layer boundary-zone[31]. The microstructure of the upper region of the deposited material changed from fine dendrite grains to the equiaxed grains which remained the finest structure. The tensile strength was about 295MPa, with an elongation of about 36%.

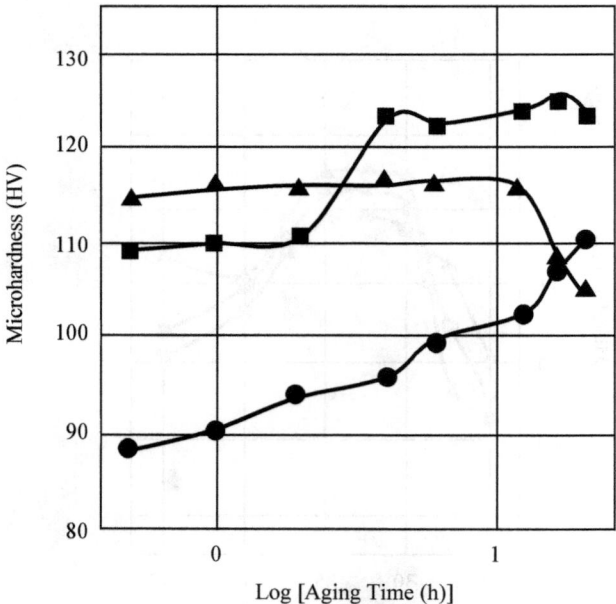

*Figure 10. Microhardness of selective laser melt (scan speed: 800mm/s) Al-3.02Mg-0.2Sc-0.1Zr as a function of aging time at temperatures of 375C (triangles), 325C (squares) or 275C (circles)*

Selective laser melt printed Al-3.02Mg-0.2Sc-0.1Zr alloy with a relative density of 99.2% was obtained[32] by optimising the laser parameters. Following aging, the hardness had improved to 120HV from the 85HV of as-prepared material. Increasing the aging temperature and time could initially improve the hardness, but excessively high aging temperatures and times decreased the hardness (figures 9 and 10). The above change in hardness was achieved by heating at 300 to 350C for 4 to 12h. A maximum tensile strength of 400MPa and yield stress of 327MPa could be obtained. Nano-sized $Al_3(Sc,Zr)$ precipitates at the grain-boundaries hindered grain growth, thus maintaining the grain size during aging.

Selective laser melting was used[33] to prepare samples of high-performance Al-Mg-Sc-Zr alloy. Small amounts of spherical $Al_3(Sc,Zr)$ nanoprecipitates were embedded at the bottom of the melt pool at low scan-speeds. No precipitates were present in the matrix, when using a relatively high scan-speed, due to the combined effects of variations in the Marangoni convection vector, the ultra-short lifetime of the liquid and the rapid cooling rate. As a result, an increased hardness of $94HV_{0.2}$ and a lower wear-rate of $1.74 \times 10^{-4}mm^3/Nm$ were obtained when a much lower scan-speed was used.

During the additive manufacture aluminium alloys, voids and cracks can spoil the product. Laser powder-bed fusion of AA6061 was performed[34] at 500C in an attempt to produce crack-free components. Melt-pool banding, a common result of the process, was eliminated. Micro-indentation and tensile testing of as-prepared samples indicated an average hardness of 54HV, together with a yield stress, ultimate strength and elongation of 60MPa, 130MPa and 15%, respectively. These values were comparable to those for annealed and T6 heat-treated wrought products. Columnar grain growth occurred in the build direction, with the as-prepared powder-bed heated product being characterized by a [100] texture, with elongated (25μm by 400μm) grains plus intragranular and intergranular non-coherent Al-Si-O precipitates. Driven by a desire to weld steel and aluminium in the solid state, 9kW ultrasonic additive manufacturing has been used to produce AA6061-4130 steel combinations[35]. Push-pin testing showed that steel–aluminium joints failed across multiple layers while Al–Al joints delaminated from the substrate. The change in failure morphology was due to the formation of metallurgical bonding in the case of Al–steel constructions. The texture was identical at the interfaces of Al–steel, Al–Al and Al–Ti joints, indicating that bond formation in every case depended upon plastic deformation across several materials. No changes in the bonding mechanism occurred when the materials which were used as foils and substrates were interchanged.

*Al-Mn*

The stresses generated by pre-strained shape-memory alloy NiTi during heating have been used to control the coefficient of thermal expansion of metal-matrix composites[36]. The latter consisted of an AA3003-H18 matrix, containing embedded NiTi ribbons, which had been created by means of ultrasonic additive manufacturing. The parameters which permitted control of the overall thermal expansion were the NiTi volume fraction and the degree of pre-strain of the NiTi. Mechanical coupling between the matrix and ribbons involved mainly friction, and the shear strength of the interface was 7.28MPa. The coefficient of thermal expansion of the composite could be caused to be zero at 135C. Ultrasonic additive manufacturing is based upon ultrasonic metal welding. Composites which are created by using this process can be exposed to temperatures as low as 25C; as compared with the temperatures of at least 500C which are involved in the use of other methods. Work was also carried out[37] on optimizing the welding conditions for AA6061 components in order to guide the design of ultrasonic additively manufactured parts containing smart materials.

*Al-Si*

Samples of Al-12Si having various layering patterns were prepared[38] by using selective laser melting. Samples with differing layer patterns had similar crystallite sizes and aluminium lattice parameters and comparable amounts of free residual silicon, but differing degrees of texture. The yield stress varied from 235 to 290MPa and the ultimate tensile strength from 385 to 460MPa. The ductility was between 2.8 and 4.5%. When base-plate heating was used, the resultant materials had an improved plasticity: base-plate temperatures of 473, 573 and 673K led to ductilities of 3.5, 3 and 9.5%, respectively. Overall, the room-temperature yield stress, ultimate tensile strength and ductility could be varied over the ranges, 115 to 290MPa, 220 to 460MPa and 2.8 to 9.5%, respectively, during the selective laser melting process, thus permitting their tailoring in order to match a specific application.

Bulk samples of $TiB_2$-reinforced aluminium-silicon alloy have been prepared[39] by using tungsten inert gas wire and arc additive manufacturing methods in which 1.6mm-diameter filler wire was used as the deposition metal, followed by T6 heat treatment. The texture of the original samples, parallel and perpendicular to the weld direction, was similar and consisted of columnar dendrites and equiaxed crystals. Following the T6 heat treatment, the hardness had increased from 62.83 to 115.85HV, the yield stress was 273.33MPa, the average tensile strength was 347.33MPa and the average elongation-to-fracture was 7.96%. The fracture mode was ductile, in spite of the presence of porosity.

A semi-solid continuous micro fused-casting additive method has been developed[40] for producing ZL101 (Al-7wt%Si) alloy strip in which a semi-solid metal slurry is pressed from an outlet in the bottom of a crucible to a movable plate. Cooling is provided by the movement of the substrate in the micro fused-casting area. By using 3-dimensional control software, the alloy strip is solidified and forms layer-by-layer. The product has a uniform structure when the movement speed is 20mm/s and the temperature is 590C, giving an ultimate tensile strength and elongation of 242.59MPa and 7.71%, respectively, with an average hardness of 82.55HV.

The controlled short circuit metal inert gas method has been used[41] to free-form simple shapes from AA4047 (Al-12Si) alloy. This could produce components comprising fine primary aluminium dendrites having an arm-spacing ranging from 2.0 to 4.8µm. This process could refine the silicon eutectic phase to a fibrous morphology because the solidification front velocity[42] exceeded a critical value of 0.9mm/s. The hardness was constant at 70HV throughout the height of the free-formed component. The bending strength was similar to that of as-cast components, being 388MPa for the former and 357MPa for the latter. Meanwhile the ductility doubled, from 9 to 18%, and this was associated with the presence of the above fibrous eutectic instead of the flaky morphology which was observed in cast samples. Micro laser metal wire deposition was used[43] to prepare aluminium-alloy samples from 0.4mm AA4047 wire. Thin sub-mm walls, free from pores and cracks could be obtained. The product exhibited an homogenous microhardness profile over the build direction, with a value of about 120HV. The improved properties were attributed to redistribution of the silicon in the form of a micrometric network within the aluminium matrix along the deposit.

A study was made of the effect of laser additive manufacturing on the properties of Al–12wt%Si alloy, with and without a $TiH_2$ foaming agent[44]. The porosity varied from an initial level of 20.9% in the absence of a foaming agent, to 32.3 and 45.9% for additions of 5 and 10% of foaming agent, respectively. The average microhardness ranged from 100 to 130HV, and the shape of the stress–strain compression curve was the same as that found when using powder metallurgy and casting methods.

A study of cold gas dynamic spray-produced A380 (Al-8.5Si-3.5Cu-3wt%Zn) silicon-particle reinforced composites, hot-rolled to various reductions, has suggested[45] that hot-rolled samples have an improved strength and ductility. Specimens with a thickness reduction of 40% had the highest ultimate tensile strength (420MPa) and elongation (5%), as compared with the equivalent values (100MPa, 0%) for as-sprayed and conventionally heat-treated samples (186MPa, 0.93%). The improvement in properties was attributed mainly to the elimination or reduction of interlayer defects, together with *in situ* composite microstructure formation. Such a composite structure formed via the progressive refinement and uniform distribution of silicon particles in an α-aluminium matrix containing coherent θ′ precipitates.

Selective laser melting was used to prepare Al-10Si-0.4%Mg alloy, and specimens having a relative density of almost 100% were obtained[46]. The ultimate tensile strength was greater than 450MPa and the elongation at fracture was greater than 10%. These values were much higher, than those of high-pressure die-cast material, because of the almost 100% relative density and sub-micron dendrite-cell microstructure. Following T6 heat-treatment, the elongation of the laser-melted material had increased by about 20% but the ultimate tensile strength had decreased.

*Table 3. Properties of selective laser melted AlSi10Mg following stress-relief treatment*

| Build Direction | Density (kg/m$^3$) | E (GPa) | G (GPa) | Poisson Ratio |
|-----------------|--------------------|---------|---------|---------------|
| vertical        | 2648               | 77.8    | 29.2    | 0.33          |
| horizontal      | 2637               | 77.1    | 28.9    | 0.33          |

A study of the properties of three different AlSi7Mg powders showed[47] that the presence of fine particles increased the pick-up of moisture, thus increasing the total particle surface-energy and interparticle cohesion. This hindered powder-flow and therefore the spreading out into the uniform layers which are required for optimum printing. When spherical particles which are larger than 48μm, with a narrow particle distribution, are present the moisture absorption, surface energy and cohesion decrease; thus increasing spreadability.

*AlSi10Mg*

A study was made[48] of the effects of pre-heating, during the selective laser melting of aluminium components, upon thermally-induced distortion. A marked reduction in distortion, as compared to the distortion without pre-heating, began for a pre-heating temperature of 150C. At a pre-heating temperature of 250C, distortions could no longer be detected. In addition to reducing distortion, pre-heating avoided the risk of stress-related cracks appearing in a component. It was also noted that the observed hardness, of 90HV$_{0.1}$, at a pre-heating temperature of 250C exceeded the minimum specified hardness of die-cast AlSi10Mg parts.

The mechanical properties of selective laser melt produced AlSi10Mg specimens were investigated[49] at 25 to 400C following stress relief (table 3). The yield stress markedly decreased, and the elongation increased, at temperatures above 200C (table 4). The ultimate tensile stress continuously decreased with temperature. The stress exponent and apparent activation energy for creep (table 5) were 25 and 146kJ/mole, respectively. The plastic deformation during creep was controlled by dislocation movement in the primary aluminium grains, and the structure was essentially that of a composite reinforced with sub-micron silicon particles. Not surprisingly, the creep resistance was close to that for aluminium-matrix particle-reinforced composites.

*Table 4. Properties of selective laser melted AlSi10Mg as a function of temperature*

| Temperature (C) | Yield Stress (MPa) | UTS (MPa) | Elongation (%) |
|:---:|:---:|:---:|:---:|
| 25 | 204 | 358 | 7.2 |
| 50 | 198 | 341 | 8.5 |
| 100 | 181 | 286 | 10.0 |
| 150 | 182 | 241 | 14.7 |
| 200 | 158 | 189 | 16.4 |
| 250 | 132 | 149 | 30.9 |
| 300 | 70 | 73 | 41.4 |
| 350 | 30 | 33 | 53.8 |
| 400 | 12 | 14 | 57.4 |

The selective laser melting process tends to result in anisotropic material behavior. Studies of AlSi10Mg revealed[50] marked directional dependences. The Young's modulus varied from 62.5 to 72.9GPa, with the Poisson ratio fluctuating between 0.29 and 0.36. The ultimate tensile strength ranged from 314 to 399MPa, with elongations to failure ranging from 3.2 to 6.5% in the non heat-treated condition.

A theoretical model, which linked the microstructure to processing parameters such as the laser-power, scanning-speed and pre-heating, was used to estimate the texture evolution as a function of number of layers of AlSi10Mg when the scanning direction was unidirectional and there was no rotation of the scanning direction between layers. It was predicted that the texture would attain a steady state after five layers[51].

*Table 5. Creep properties and microhardness of AlSi10Mg*

| Temperature (C) | Stress (MPa) | Strain Rate (/s) | Lifetime (min) | Hardness (GPa) |
|---|---|---|---|---|
| 225 | 147 | $1.06 \times 10^{-4}$ | 9 | 0.94 |
| 225 | 137 | $1.47 \times 10^{-5}$ | 32 | 0.91 |
| 225 | 127 | $1.78 \cdot x \ 10^{-6}$ | 360 | 0.90 |
| 225 | 117 | $3.14 \cdot x \ 10^{-7}$ | 3944 | 0.88 |
| 250 | 117 | $4.01 \cdot x \ 10^{-6}$ | 282 | 0.89 |
| 275 | 117 | $1.21 \cdot x \ 10^{-5}$ | 33 | 0.91 |
| 300 | 117 | $3.50 \cdot x \ 10^{-5}$ | 8 | 0.92 |

*Table 6. Ultimate tensile strength of laser direct metal deposited AlSi10Mg*

| Scanning speed (mm/min) | UTS (MPa) |
|---|---|
| 720 | 125 |
| 600 | 161 |
| 480 | 162 |
| 0 (cast) | 145 |

A comparison of AlSi10Mg alloys which had been prepared by using additive manufacturing or powder metallurgy methods showed that the former contained coarser aluminium grains, but much finer silicon precipitates[52]. As a result, the additively manufactured alloy had a more than 100% greater strength and hardness than that of the powder-metallurgy alloy, due to the ultrafine silicon, while retaining an acceptable ductility.

Laser direct metal deposition of AlSi10Mg showed[53] that, with decreasing scanning speed, the microhardness of the product decreased while the tensile properties improved and indeed exceeded those of cast material (table 6). The microstructure differed in the various parts of the deposit: the bottom consisted mainly of cellular crystals, while the middle part comprised mainly columnar crystals and the upper part was made up mainly of equiaxed crystals. There were consequent changes in the microhardness with increasing deposition height, in that the middle part (columnar crystals) had a higher microhardness while the other parts had a lower microhardness. The microhardness was affected by the traverse speed, with a high speed leading to a greater microhardness. Reducing the scanning-speed from 720 to 480mm/min could improve the tensile properties. Porosity and powder remnants could affect the tensile properties.

*Table 7. Properties of selective laser melted AlSi10Mg macroscopic structures*

| Cell (mm) | Strut (mm) | Form | E (MPa) | UTS (MPa) | $\sigma_{0.2}$(MPa) |
|---|---|---|---|---|---|
| 4 | 1 | bcc | 271 | 10 | 9 |
| 4 | 1.2 | bcc | 745 | 23 | 17 |
| 5 | 1 | bcc | 65 | 4 | 4 |
| 5 | 1.2 | bcc | 299 | 10 | 8 |
| 4 | 1 | bccz | 1139 | 29 | 19 |
| 4 | 1.2 | bccz | 1721 | 54 | 36 |
| 5 | 1 | bccz | 720 | 15 | 12 |
| 5 | 1.2 | bccz | 1146 | 29 | 21 |

Looking at more macroscopic structures, trabecular assemblies have been made[54] from AlSi10Mg, by means of selective laser melting, in which the cell size was 4 or 5mm and the strut size was 1 or 1.2mm. One shape, termed bcc (figure 11), was essentially a 2-dimensional diamond lattice while the other, termed bccz (figure 12), was the same but with vertical struts. The most influential factor in mechanical performance was the type of cell (figures 13 and 14). For a given geometry, the relative density determined the behaviour; a greater density leading to a better performance (table 7). All of the specimens failed due to fracture at 45° to the compression axis. Structures with struts remained united while the others separated. In the case of cells with vertical struts, the failure mode was most affected by the size of the cell. Those with a size of 5mm collapsed due to buckling of the vertical struts, with no deformation of the remainder of the specimen. The 4mm cells also collapsed due to buckling of the vertical struts, but only after having absorbed a lot of energy in deformation of the entire specimen. Thus 4mm cells responded better to compressive and impact loads, and collapse was more predictable. In the case of cells without vertical struts, sample collapse was due to folding and detachment at the nodes. The 5mm cells fractured under little deformation, while 4mm cells withstood greater deformation.

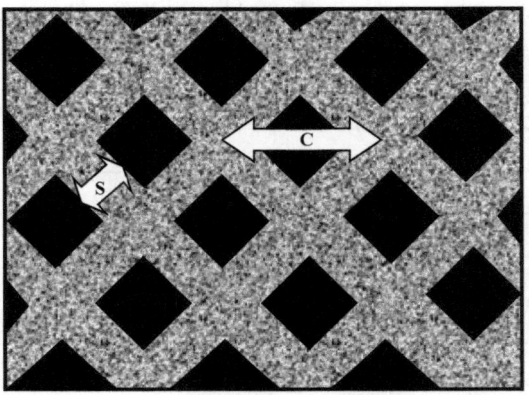

*Figure 11. Form of bcc structure: C, cell size, S, strut size*

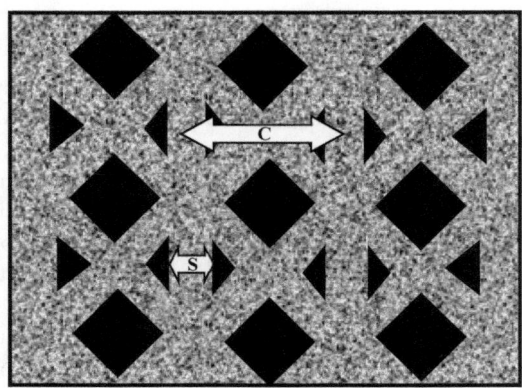

*Figure 12. Form of bccz structure: C, cell size, S, strut size*

A statistical approach has been used[55] to determine the effect of powder re-use upon the properties and the surface quality of additively manufactured AlSi10Mg parts. Such re-use had some statistical significance in terms of the yield stress and ultimate tensile strength, but the static and cyclic mechanical properties of the product nevertheless exhibited little variability and remained within specifications.

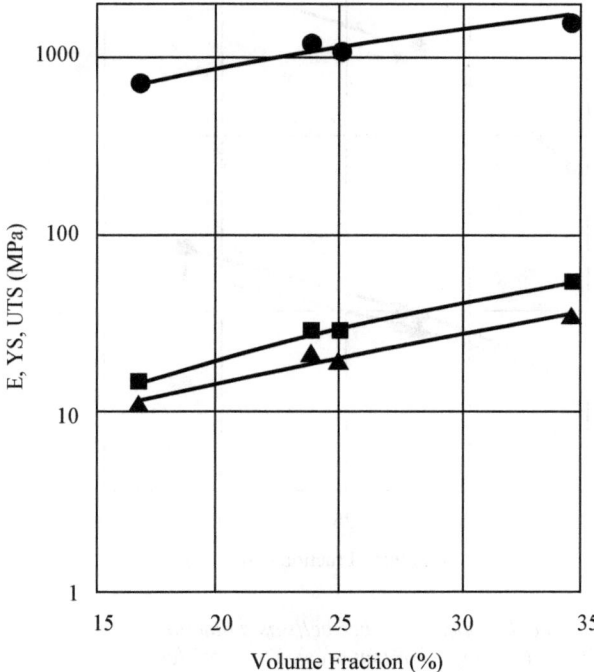

*Figure 13. Mechanical properties of bccz cells as a function of volume fraction
E (circles), UTS (squares), YS (triangles)*

Solutionizing and aging treatments were carried out[56] on samples prepared by direct laser sintering or gravity-casting. The results were related to the microstructure and degree of porosity. In the case of the additively manufactured samples, heat treatment could greatly impair the performance of the alloy due to an increase in porosity resulting from gas which was entrapped during deposition. The higher the solutionizing temperature, the greater was the increase in such defects. No similarly marked effect was observed in the case of conventionally cast material.

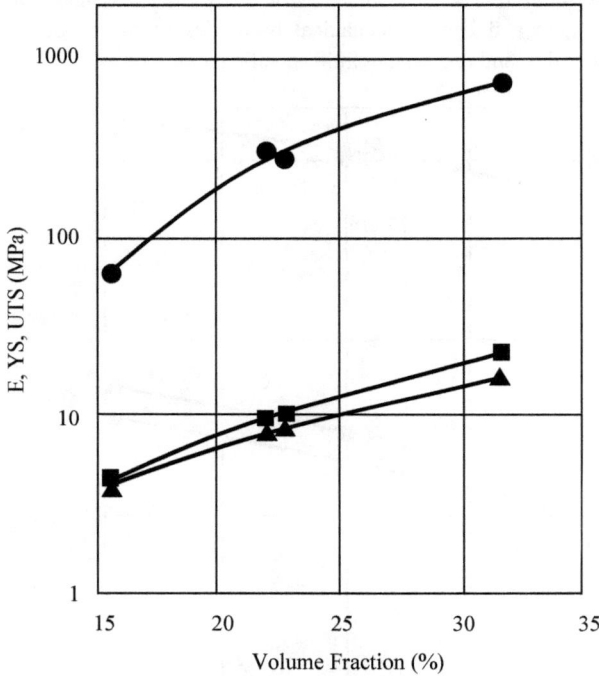

*Figure 14. Mechanical properties of bcc cells as a function of volume fraction E (circles), UTS (squares), YS (triangles)*

Selective laser melting could produce composites with a matrix of AlSi10Mg and SiC reinforcement[57]. The reinforcing phases included non-melted micron-sized SiC particles, *in situ* formed micron-sized $Al_4SiC_4$ strips and *in situ* produced sub-micron $Al_4SiC_4$ particles. With increasing laser energy density, the extent of *in situ* reaction between SiC particles and the matrix increased and resulted in more $Al_4SiC_4$ reinforcement being formed. The densification-rate of the composites increased as the applied laser energy density was increased. An approximately 96% theoretical density could be obtained by using laser linear energy densities greater than 1000J/m. Due to the presence of multiple reinforcing phases, the microhardness was $214HV_{0.1}$, with a relatively low coefficient-of-friction of 0.39 and a wear-rate of $1.56 \times 10^{-5}mm^3/Nm$. At very high laser energy inputs, the grain size of the *in situ* formed strip or particulate $Al_4SiC_4$ reinforcement greatly increased. Such grain-coarsening, and the formation of microscopic interfacial shrinkage porosity, impaired the mechanical properties of the composites.

The effect of T6-type heat-treatment, following the selective laser melt preparation of AlSi10Mg, was investigated[58]. Nano-indentation tests indicated a uniform hardness of the as-processed material, while marked spatial variations were found following heat treatment. As a result of phase transformations. The microhardness of the as-processed material was greater than that of the equivalent die-cast material. The use of heat-treatment softened the material; reducing the microhardness from 125 to 100HV. The ultimate tensile strength of 333MPa was reduced by 12% due to heat treatment, but there was a threefold increase in the strain-to-failure (table 8). These properties again exceeded those of die cast samples. A markedly high compressive yield strength was exhibited by as-processed material, in that the compressive yield strength was 370GPa and the compressive strength at 25% strain was 714MPa.

*Table 8. Tensile properties of selective laser melt processed AlSi10Mg*

| Condition | 0.2%YS (MPa) | E (GPa) | UTS (MPa) | Elongation (%) |
|-----------|--------------|---------|-----------|----------------|
| As-processed | 268 | 77 | 333 | 1.4 |
| T6 heat-treated | 239 | 73 | 292 | 3.9 |

Samples of AlSi10Mg were produced[59] by using a direct energy deposition process. Columnar dendritic structures predominated, although there was a cellular morphology near to the substrate and equiaxed dendrites at the top of the deposit. The morphology at the melt pool boundary differed from that in the core of the layers. The dendrites were unidirectionally oriented at an angle of about 80° to the substrate. The region which exhibited a cellular morphology had the greatest hardness value (65HV), while the uppermost section of the deposit, with an equiaxed dendritic structure, had the lowest hardness (53HV). This weakness was attributed to the coarser microstructure.

AlSi10Mg reinforced with TiC particles was deposited onto AA5052 aluminium alloys by means of laser additive manufacturing[60]. Use of a suitable laser-energy input produced a uniform distribution of needle-like TiC particles which precipitated from the grain boundaries and led to improved mechanical properties. The optimum process parameters were a laser-power of 2500 to 3500W and a scanning velocity of 600mm/min. At lower or higher laser-energy inputs, aggregates formed in the coatings. When using the optimum processing parameters, the maximum microhardness and tensile strength of the AlSi10Mg-TiC alloys were 139.1$HV_{0.05}$ and 278.8MPa, respectively. The large numbers of dimples on the fracture surfaces indicated the occurrence of ductile fracture.

*Table 9. Properties of Al-6.8Zn-6.5Si-2Mg-1.3Cu in various conditions and orientations*

| Condition | Orientation | Yield stress (MPa) | UTS (MPa) | Elongation (%) |
|---|---|---|---|---|
| as-prepared | vertical | 313 | 386 | 2.2 |
| T5 temper | vertical | 370 | 432 | 1.4 |
| As-prepared | horizontal | 332 | 447 | 2.3 |
| T5 temper | horizontal | 402 | 449 | 1.32 |

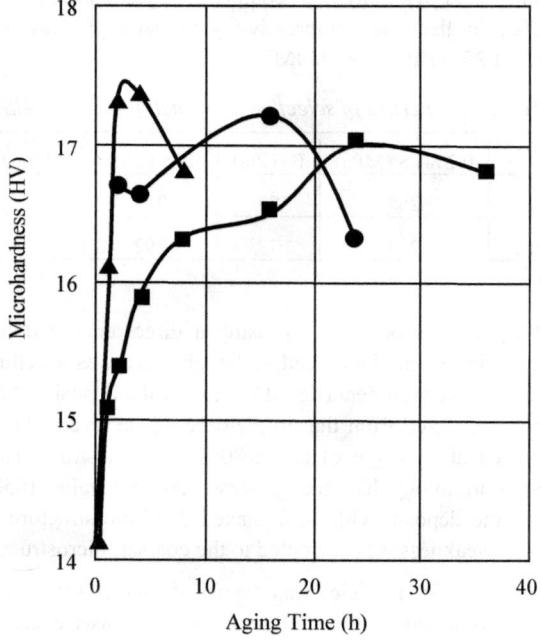

*Figure 15. Effect of aging Al-6.8Zn-6.5Si-2Mg-1.3Cu at 135C (squares), 150C (circles) and 165C (triangles)*

Micro-lattices of AlSi10Mg having various unit-cell structures, numbers of unit cells and strut diameters have been made[61] by means of selective laser melting. They had a maximum specific compressive strength of 83.113MPacm$^3$/g. This was much higher than the magnitude expected for most metallic and non-metallic micro-lattices. Four different failure modes occurred during compression testing: contact-region crushing, consecutive diagonal cracking at 45° to the loading direction, elastic/plastic buckling and plastic deformation, and single diagonal cracks at 45° to the loading direction. Micro-lattices having differing unit cells exhibited differing strength-density relationships, due to the variety of failure modes.

## Al-Zn

Laser-based additive manufacturing is not generally suitable for preparing the AA7050 alloy because of potential loss of the latter's low boiling-point constituents, such as magnesium and zinc. One solution is to use AA7050 powder which is coated with nickel. The resultant deposits are free from defects such as porosity and incomplete fusion, but the nickel tends to segregate in the interdendritic boundaries and form the brittle Al$_3$Ni phase. As-deposited nickel-coated AA7050 may therefore possess almost no ductility in tension. When laser-deposited samples were friction-stir processed[62] in order to refine and redistribute the Al$_3$Ni particles in the α-aluminium matrix however, tensile testing indicated a satisfactory combination of properties, with the yield stress being 178MPa, the ultimate tensile strength 302MPa and the elongation being 6%. There was a microstructural change, from columnar to equiaxed dendrites, in going from the bottom to the top of each deposited layer.

Selective laser melting was used[63] to process Al-6.8Zn-6.5Si-2Mg-1.3Cu alloy so as to have a relative density of 99.8%. The presence of silicon had positive effects on the alloy. The formation of eutectic during the final stage of solidification improved the liquid feeding of intergranular regions. It was deduced that the solidification and cooling rates during selective laser melt processing were sufficiently rapid to produce supersaturated solid solutions having good aging capabilities. Additional solution-treatments could even be counter-productive. In the T5-tempered state, the hardness was 174HV (table 9). A good response to direct aging (figure 15), from the as-processed state, led to a yield strength and ultimate tensile strength of 402 and 449MPa, respectively.

Billets of AA7075 alloy have been made[64] by using the micro-droplet deposition technique. Three sets of samples were prepared under differing temperature conditions, showing that the temperature of the metal droplets and substrate markedly affected the resultant tensile strength. By using optimum temperature settings, the billets could attain a tensile strength of 373MPa and an elongation of 9.95%; figures which were very similar to those of extruded material. There was an appreciable change in the failure mechanism with improving metallurgical bonding.

Due to a persistent uncertainty as to how best to catalogue members of the new material group known as 'high-entropy alloys', where no particular element can be said to be the base element, some examples are listed here for purely alphabetical reasons even though their characteristics are closer to those of a superalloy than to those of an aluminium alloy.

In a study[65] of laser metal deposited samples of the high-entropy alloy, AlFeCoCrNi, the ratio of aluminium to nickel was decreased in order to produce a transition of the solid solution structure from body-centred cubic to face-centred cubic. The lattice parameter then increased from 288 to 357nm and the hardness decreased from 670 to 149HV. Differential thermal analysis was used[66] to track the solidification paths of each of the compositions. Annealing at high temperatures led to $\sigma$-phase transformation in some of the compositions which, together with solid-solution strengthening, explained the high softening-resistance of the alloys. Laser engineered net shaping processed compositionally-graded $AlCrFeMoV_x$ high-entropy alloy, deposited from a blend of elemental powders, has recently been studied[67]. The composition ranged from 0.3 to 18.5at%V over a distance of about 20mm. A single solid-solution body-centered cubic structure existed over the whole composition range. The high solubility of vanadium permitted variable degrees of solid-solution strengthening of the structurally simple body-centred cubic matrix. The hardness increased from 485 to 581HV as the vanadium content was varied from 0.3 to 18.5at%.

## Cobalt

Additive manufacturing is a particularly good solution to the problem of making complicated components out of notoriously intractable materials. Laser beam melting has therefore been applied[68] to the forming of commercially available tungsten carbide powder with a 17% cobalt content, yielding bending strengths greater than 1GPa and a fracture toughness ranging from 10 to 12MPa$\sqrt{m}$. Cobalt-carbide composites could be produced by gel-printing using hydroxyethyl methacrylate loaded with 47 to 56vol% of WC-20Co solid and then sintering under vacuum. The printed product had a good shape, was precise and possessed an homogeneous microstructure. The best sintered material[69] had a transverse rupture strength of 2612.8MPa. It is feasible to create 3-dimensional cermets, via selective laser melting, by using 5 to 35$\mu$m boron carbide particles surrounded by 2$\mu$m cobalt-based layers[70]. High (37%) porosity homogeneous structures containing boron carbide grains with a hardness of 2900 to 3200HV, embedded in a cobalt-based matrix, can be obtained. New phases also form as a result of interaction of the $B_4C$ with the cobalt-based matrix.

Specimens of cobalt-chromium-molybdenum and cobalt-chromium-tungsten alloys were prepared using direct metal printing and laser-based methods respectively, and tested at 0, 30, 60 and 90° inclinations in both the as-processed and heat-treated conditions. This revealed[71] that improved tensile properties and hardness values followed a short heat-treatment of the Co-Cr-Mo alloy, although the microstructure was not homogenised. Annealing of the Co-Cr-W alloy led to an increased modulus and elongation (apart from the 90° orientation), but with a decreased yield strength, ultimate tensile strength and hardness in comparison with those of the as-processed material. In similar work[72], cobalt-chromium-molybdenum parts were prepared using various additive manufacturing techniques. Fully-dense parts were produced by selective laser melting and direct metal laser sintering. The relative density of parts produced using either method was 99.3%, and use of a higher energy-density could potentially effect a higher sample density in both cases. The average roughness of the upper surfaces of the former parts was 3.4μm, as compared with 2.83μm for direct metal laser sintered parts. The roughness of the side-surfaces of selective laser melted was 4.57μm, as compared with 9.0μm for direct metal laser sintered material. The higher roughness values for the side-surfaces as compared with the upper surfaces in both cases was attributed to balling (particle coalescence) effects.

*Table 10. Mechanical properties of binder-jetted copper*

| Powder | Condition | $\rho$(%) | Porosity (%) | UTS (MPa) | Elongation (%) | E (MPa) |
|--------|-----------|-----------|--------------|-----------|----------------|---------|
| 25μm | sintered | 77.7 | 22.3 | 82.0 | 28.9 | 283.7 |
| 25μm | HIPed | 82.4 | 17.6 | 129.3 | 61.8 | 209.2 |
| 17μm | sintered | 83.6 | 16.4 | 115.8 | 48.5 | 238.8 |
| 17μm | HIPed | 85.8 | 14.2 | 135.3 | 57.8 | 234.1 |
| bimodal | sintered | 90.5 | 9.5 | 144.9 | 43.9 | 330.1 |
| bimodal | HIPed | 97.3 | 2.7 | 176.4 | 67.2 | 262.5 |

A non-beam approach to the additive manufacturing of high-entropy alloys has recently been developed[73] which is based upon the 3-dimensional extrusion of inks containing a blend of oxides, $Co_3O_4$ - $Cr_2O_3$ - $Fe_2O_3$ - NiO, in nanopowder form. These were then reduced in $H_2$ to the respective metal, interdiffused and sintered to give near-full density face-centered-cubic equi-atomic CoCrFeNi. This material could be used to create micro-lattices, having strut diameters as small as 100μm, which possessed excellent mechanical properties at room and cryogenic temperatures.

Electron-beam melting of Co-26Cr-6Mo-0.2C powder was used[74] to make simple solid components with a density of 8.4g/cm$^3$ and reticulated mesh material with a density of 1.5g/cm$^3$. Orthogonal melt rastering of the electron-beam during the additive manufacturing process produced orthogonal $Cr_{23}C_6$ precipitate arrays, with dimensions of about 2μm, in the plane perpendicular to the build direction while carbide columns formed in the vertical plane parallel to the build direction. The micro-indentation hardness ranged from 4.4 to 5.9GPa, corresponding to a yield stress and ultimate tensile strength of 0.51 and 1.45GPa, with the elongations ranging from 1.9 to 5.3%. Annealing treatments produced an equiaxed face-centred cubic grain structure with some grain-boundary carbides, copious annealing twins and often a high density of intrinsic {111} stacking faults within the grains. The reticulated mesh microstructure consisted of dense carbide arrays having an average micro-indentation hardness of 6.2GPa; some 25% higher than that of the fully dense products.

Laser engineered net shaping was used[75] to deposit CoCrMo onto a metallic substrate. The microstructures of the deposits were uniform and fine. Coatings which were produced by using a high (350W) laser-power, low (5g/min) powder feed-rate and high (20mm/s) scan velocity had the greatest hardness (446HV) and wear resistance (1.80mm$^3$/Nm). On the other hand, the corrosion resistance was high for deposits produced by using a low (200W) laser-power, low (5g/min) powder feed-rate and low (10mm/s) scan velocity.

**Copper**

The direct printing of 3-dimensional nanoscale metallic wires having the conductivity of copper, and a controlled microstructure, would obviously be useful for application in integrated circuits and flexible electronics. For this and other purposes, copper nanoparticles can be prepared in powder and ink form, and the method of synthesis can have a marked effect upon the sintering temperature and sintering quality. Surface coatings and surfactants can help to reduce agglomeration, prevent oxidation and restrict sintering of the nanoparticles at lower temperatures. The coatings improve the packing density and increase the temperature required for necking to occur, thus leading to better sintering. Most of the surface coatings were removed during sintering, leaving the sintered products with a much higher copper percentage than that of the original nanoparticles. At temperatures near to the melting point of copper particles, the surface coatings could begin to graphitize and hinder nanoparticle fusion. Thus the optimum sintering conditions for copper nanoparticle ink require a temperature high enough to break down the surface coating but low enough that copper melting and coating graphitization do not occur.

A 5kW ultrasonic generator was used[76] for the ultrasonic preparation of T2 copper foil, with an external heat source being used to optimize the temperature distribution. The solid-phase bonding strength (24.42N/mm) of the product was optimum when the heating temperature was 100C. It first increased and then decreased with increasing interfacial temperature.

*Table 11. Conductivities of binder-jetted copper*

| Powder | Condition | $\rho$(%) | Porosity (%) | Thermal (W/mK) | Electrical ($10^7$S/m) |
|--------|-----------|-----------|--------------|----------------|------------------------|
| 25µm | sintered | 77.7 | 22.3 | 245.7 | 3.0 |
| 25µm | HIPed | 82.4 | 17.6 | 256.5 | 3.7 |
| 17µm | sintered | 83.6 | 16.4 | 262.3 | 3.8 |
| 17µm | HIPed | 85.8 | 14.2 | 266.3 | 3.7 |
| bimodal | sintered | 90.5 | 9.5 | 293.5 | 4.7 |
| bimodal | HIPed | 97.3 | 2.7 | 327.9 | 5.2 |

Binder jetting, for which the main problem is to produce fully-dense homogeneous components without requiring infiltration, was applied to copper.[77] Parts having porosities ranging from 2.7 to 16.4% were produced by varying the powder morphology and the post-processing sintering and hot isostatic pressing conditions. Those parts having the lowest porosity had a tensile strength of 176MPa (table 10). This was equal to 80.2% of the strength of wrought material. The thermal conductivity of 327.9W/mK (table 11) was equal to 84.5% of that of wrought material.

Laser additive manufacturing has been used[78] to produce copper-graphene nanocomposites. The addition of graphene markedly improved the mechanical properties, with nano-indentation tests showing that the average modulus and hardness of the composites were 118.9GPa and 3GPa, respectively. These represented improvements of 17.6 and 50% when compared with the equivalent figures for pure copper.

*Table 12. Tensile properties of wire-arc additively manufactured Cu-9at%Al*

| Condition | UTS (MPa) | 0.2%YS (MPa) | Elongation (%) |
|---|---|---|---|
| as-prepared | 231 | 63 | 63 |
| annealed (900C, 2h) | 241 | 67 | 76 |

The use of electron-beam melting to create copper components by additive manufacture, using low-purity atomized copper powder containing a high density of $Cu_2O$ precipitates, produced[79] interesting precipitate-dislocation interactions. These usually consisted of equiaxed precipitate-dislocation 1 to 3μm cell-like arrays in the horizontal reference plane perpendicular to the build direction, with elongated or columnar-like arrays that were from 12 to more than 60μm in length. The associated hardness ranged from 83 to 88HV, as compared to the original copper-powder micro-indentation hardness of 72HV and the 57HV hardness of commercial copper plate.

Wire-arc additive manufacturing was used[80] to deposit Cu-9at%Al onto pure copper plates by separately feeding pure copper and aluminium wires into the melt pool generated by gas tungsten arc welding. The as-prepared material was further treated using homogenization heat-treatment at 800 or 900C (table 12). With increasing annealing temperature, the precipitate phase content decreased and there were gradual improvements in strength and ductility, with little change in the microstructure.

Just as zinc alloys pose a problem, for selective laser melt processing, because of losses due to evaporation, so too do copper alloys cause difficulties due to their high thermal conductivity and poor laser absorption. These difficulties have been explored[81] with regard to the formation of Cu-4Sn. Single-track experiments were used to identify possible processing-parameter windows. A Latin-square approach, with orthogonal parameter arrays, was used to isolate various parameters, and analysis-of-variance was used to find statistical relationships which described the effects of laser-power, scanning-speed, etc., upon the relative density and Vickers hardness. It was found that the relative density and hardness of Cu-4Sn were controlled mainly by the laser power. The maximum relative density was better than 93% of theoretical and the maximum hardness was 118HV. The best tensile strength, of 316 to 320MPa, was less than that of pressure-processed Cu-4Sn but might be improved by reducing the number of defects.

Specimens of Cu-Cr-Zr-Ti alloy (0.50 to 0.70Cr, 0.02 to 0.05Zr, 0.02 to 0.05wt%Ti), having an initial relative density of 97.9%, were prepared[82] by using the selective laser melting technique. The resultant microstructure comprised grains which were elongated along the build direction, with sizes ranging from 30 to 250μm. Mechanical testing showed that the material had an ultimate tensile strength of 195 to 211MPa and an elongation-to-failure of 11 to 16% at 20C. Samples which were taken parallel to the build direction had slightly higher ultimate tensile strengths and elongations, as compared with those taken perpendicular to the build direction (table 13). The ultimate tensile strength of the selective laser melt samples was however some 20 to 25% lower than that of hot-rolled samples.

*Table 13. Properties of selective laser-melted Cu-Cr-Zr-Ti specimens*

| Orientation | Temperature (C) | UTS (MPa) | Elongation (%) |
|---|---|---|---|
| perpendicular* | 20 | 195.1– 198.0 | 10.8–11.7 |
| perpendicular* | 600 | 69.5–86.2 | 4.4–5.7 |
| perpendicular* | 800 | 31.3–33.3 | 6.3–12.0 |
| parallel* | 20 | 210.0– 211.0 | 13.1–15.8 |
| parallel* | 600 | 82.2–82.3 | 4.2–7.7 |
| parallel* | 800 | 41.2–46.6 | 7.8–12.1 |

* with respect to build direction

Studies have been made of how the angle of the laser scanning pattern, also known as the build orientation, affects the products of selective laser sintering. In experiments performed on Cu-Sn-Ni metallic layers[83], the angle of the scanning pattern relative to the transverse dimension of the piece was set at 0, 30, 45, 60 or 90° while maintaining a constant scanning-speed and laser-beam power and producing specimens of various thickness. There was a noticeable effect of the scan-angle upon the mechanical strength and the degree of densification of the sintered specimens. The thickness of the resultant monolayer changed negatively with an increased scan angle while the relative density changed positively. The minimum porosity and maximum ultimate tensile strength were associated with a scan-angle of 60°.

Selective laser melting has been used[84] to prepare fully dense samples of Cu-15Sn bronze which had very fine grains with cellular and dendritic structures. Following annealing, the ultimate tensile strength varied from 661 to 545MPa, the elongation-to-fracture ranged from 7.4 to above 20% and the hardness ranged from 212.3 to $168HV_{0.3}$ (table 14). These results were superior to those of conventionally produced material. It was suggested that, as a result of annealing, the predominant strengthening mechanism had changed from Hall-Petch to solid-solution strengthening.

*Table 14. Properties of selective laser melt processed Cu-15wt%Sn*

| Condition | UTS (MPa) | 0.2%YS (MPa) | E (GPa) | Elongation (%) | $HV_{0.3}$ |
|-----------|-----------|--------------|---------|----------------|------------|
| as-processed | 661 | 436 | 90.9 | 7.4 | 212 |
| 500C, 4h | 544 | 328 | 89.1 | 20.0 | 173 |
| 600C, 4h | 545 | 328 | 87.8 | 21.2 | 168 |
| 700C, 4h | 547 | 326 | 81.5 | 25.8 | 161 |

**Iron**

A new method for the preparation of powder composites using laser additive manufacturing has been proposed[85] in which powder composites, consisting of micrometer-sized stainless-steel powder, is homogenously decorated with nano-scale $Y_2O_3$ via electrostatic deposition. The resultant specimens have superior mechanical properties at high temperatures, due to the presence of the nano-sized, homogenously distributed oxide.

Direct time-resolved images have been obtained[86] of melt-pool flow dynamics, showing that the instantaneous flow velocities in steel ranged from 0.1 to 0.5m/s. When the temperature-dependent surface tension coefficient was negative, bulk turbulence was the main flow mechanism. When the alloy had a positive temperature-dependent surface tension coefficient, surface turbulence occurred and oxides could be entrained within the resultant solid due to the higher flow velocities. Arc-type additive manufacturing routes can thus be optimised by controlling the internal melt flow.

By using the so-called big area additive manufacturing method, a thin steel wall was constructed[87] from low-carbon steel wire. The relationship between the final microstructure and the thermal history was in accord with the continuous-cooling transformation diagram of the low-carbon steel. The final microstructure depended upon the cooling rate during the austenite to ferrite/bainite transformation during initial cooling and subsequent re-heating cycles. Thus rapid cooling resulted in a small ferrite grain-size and a fine bainite structure at locations closest to the base plate. A lower cooling-rate at the side-wall, and repeated re-heating cycles into the ferrite-pearlite regions, resulted in equiaxed ferrite with medium grain size plus small amounts of pearlite. In the absence of re-heating cycles, the upper location experienced the lowest cooling-rate and this led to large-grained equiaxed ferrite and bainitic structures. A study[88] of the surface quality of wire arc additively manufactured bainitic steels showed that the optimum appearance was a smooth one having few spatters and no visible defects.

Additive manufacture can be combined with foaming-agent technology to produce steel which contains interconnected 2 to 30μm pores, giving an overall porosity of about 26vol%. Such a so-called breathable mould steel could void trapped gas during injection-moulding. A suitable choice of foaming agent, such as $CrN_x$, could produce[89] a material having a compressive strength of 1.3GPa, a microhardness of 360HV and a strain of 26%. Such values were better than those of existing conventional PM-35 breathable steel.

Binder jet 3-dimensional printing was used[90] to create steel-$TiC_x$ composites having complicated shapes. Carbide pre-forms were infiltrated with liquid 0.7wt%C steel. There was a compositional gradient in the carbide phase, with x varying from 0.7 to 0.98. Following infiltration, solidification and slow cooling, a microstructural gradient existed within the steel matrix: from ferrite in regions where the steel was in contact with carbide of low (x = 0.7) carbon content to pearlite in regions where the steel interacted with stoichiometric (x = 0.98) titanium carbide. Following heating at 900C, and oil-quenching, there was a structural gradient, from ferrite to martensite, in the steel matrix. This resulted in a hardness gradient of 700 to 1600HV.

A mild steel plus silicon-bronze bimetal was produced[91] by means of shaped metal deposition and gas metal arc welding technology. No copper was found on the steel side, but iron was present on the bronze side in the form of particles and large pieces. Silicon concentrated in the interface zone, and on the bronze side where iron appeared. The interface exhibited good adhesion, with no cracks or pores, and some degree of metallurgical bonding. The tensile strength of the bimetal attained 305MPa, and fracture occurred near to the middle of the bronze side.

Selective laser melting has been used[92] to incorporate nano-size TiC particles into pure iron, yielding 0.5wt%TiC-reinforced iron composites. When experimental data were compared with an established composite fatigue-life prediction model, it was confirmed that the TiC improved the tensile behaviour of pure iron, in that the ultimate tensile strength and yield stress were increased by 17 and 6.3%, respectively. The TiC nano-particles tended to improve the fatigue life mainly under low-cycle conditions. Any similar effect under high-cycle conditions was insignificant. This was due largely to the large degree of porosity which was introduced during the selective laser melting process.

Standard fatigue models also failed to predict accurately the fatigue life of these composites.

Selective laser melting can even bring together basically incompatible materials. Iron-manganese twinning-induced plasticity steel containing particles which were 76wt% silver – an essentially immiscible addition - could be produced[93] by means of additive manufacturing. The particles were well-distributed throughout the matrix, and hardly affected the mechanical properties, but created cathodic sites in the composite which led to an increased dissolution-rate of the alloy.

The wire arc process has been used[94] to produce iron aluminide *in situ* by separately feeding pure iron and aluminium wires into a molten pool that was generated by means of gas tungsten arc welding. The $Fe_3Al$ phase was the only one detected in the middle build-up section of the product; which accounted for most of the deposited material. The bottom of the structure contained a dilution-affected region where some acicular $Fe_3AlC_{0.5}$ precipitates existed due to carbon diffusion from the steel substrate. Microhardness data indicated the existence of relatively homogeneous material properties throughout the product. On the other hand, the tensile properties were very different in the longitudinal and normal directions (table 15), due to the epitaxial growth of large columnar grains. Other work has nevertheless shown that a functionally gradated iron-aluminium wall structure, with the aluminium composition ranging from 0 to over 50at%, could be prepared by using the wire arc additive manufacturing system. The same manufacturing system could therefore produce functionally graded iron aluminide[95].

*Table 15. Tensile properties of wire arc processed $Fe_3Al$*

| Direction | UTS (MPa) | 0.2%YS (MPa) | Elongation (%) |
|---|---|---|---|
| longitudinal | 897.8 | 810.7 | 3.5 |
| normal | 851.7 | 722.6 | 3.7 |

Wire arc additive manufacturing was used[96] to deposit thin-wall parts in 2Cr13 alloy via cold metal transfer. Pre-heating effects arising from previous layers can effectively reduce residual stresses. As an example: the cooling-rate firstly rapidly decreased and then remained stable in the $15^{th}$ to $25^{th}$ layers. Small numbers of pores were observed, together with an absence of cracks in the various layers; indicating a high degree of densification. The as-deposited microstructure consisted of martensite and ferrite, together with $(Fe,Cr)_{23}C_6$ precipitates at the α-iron grain boundaries. The martensite content increased gradually from the $5^{th}$ layer to the $25^{th}$ layer, suggesting that the metastable martensite partially decomposed into stable ferrite via carbon diffusion. The hardness and ultimate tensile strength changed slightly in the $5^{th}$ to $15^{th}$ layers and then increased rapidly from the $20^{th}$ to the $25^{th}$ layer at the expense of ductility. The fracture mode changed from ductile, in the $1^{st}$ to $10^{th}$ layers, to mixed-mode in the $15^{th}$ to $20^{th}$ layers and to brittle fracture in the $25^{th}$ layer[97]. The behaviour of 2Cr13 thin-wall parts could be tailored[98] by adjusting the inter-layer dwell-time using cold metal transfer. A short dwell-time encouraged the presence of a lower amount of γ-iron and highly-elongated ferrite grains containing ultra-fine needle-shaped martensite. This led to a slight fluctuation in microhardness. A long dwell-time led to the formation only of α-iron phase and to a periodic variation in microhardness that was related to a periodic microstructure having martensite laths within block-like ferrite. The elastic moduli of the samples were not appreciably dissimilar different[99].

A laser-driven drop-deposition technique has permitted[100] the construction of 3 adjacent tracks on a common alloy substrate sheet, the point being to compare the potential sensitisation arising from various thermal cycles. In the case of low-chromium ferritic stainless steel, a higher beam-power produced smoother tracks. The added layer was fully martensitic and had a hardness of 320HV. When the treatment-temperature peaked just below the austenitisation point, the thermal cycle arising from subsequent tracks affected the prior track via tempering. A continuous region of grain boundaries was found around the interface between the melted and heat-affected zones. In the former zone, the network became discontinuous upon approaching the surface, suggesting that the specimen was immune to sensitisation.

*AISI304*

Selective laser melting of this steel, using a layer thickness of 30mm and a speed of 70mm/s, yielded samples having a density of 92% and a hardness of 190HV[101]. At a layer thickness of 70mm, the porosity increased and cracks began to form, thus decreasing the strength and ductility. The steel remained austenitic, with no carbide films existing at the grain boundaries. This was attributed to the high melting and cooling cycles. It was noted that the cost of the process, when using feedstock powders, was less than 10% of the cost of using specialist commercial powders. There was moreover essentially no loss of product quality.

Laser melt deposition has been used[102] to produce SS304 specimens having microstructures similar to those of selective laser melted specimens. Due to differences in the grain size and phase compositions in different directions, the hardness of the laser melt deposited specimens was anisotropic. Due to a smaller grain-size, the hardness of laser melt deposited material was greater than that of conventional wrought samples. The static tensile properties of the laser melt deposited material were compared with those of conventionally produced specimens (table 16). Although the ultimate tensile strength and yield stress were lower than those of selective laser melt specimens, the elongation was greater. The fracture type was plastic. The fatigue strength of laser melt deposited material was higher (255MPa), at $10^6$ cycles (figure 16), than that of conventional wrought material, and this was attributed to the effect of grain-boundary strengthening, which hindered crack propagation.

*Table 16. Tensile properties of AISI304 produced using various processes*

| Process | UTS (MPa) | 0.2%Yield stress (MPa) | Elongation (%) |
|---|---|---|---|
| laser melt deposited | 609 | 273 | 67.3 |
| selective laser melted | 713 | 551 | 43.6 |
| conventionally wrought | 586 | 242 | 65 |
| ANSI minima | $\geq 520$ | $\geq 205$ | $\geq 40$ |

Wire arc deposition was used[103] to produce specimens of AISI304 stainless steel and ER70S mild steel. Variation of the properties occurred, in the direction of deposition and in the deposit thickness, due to variations in the local thermal history. The wear rate decreased markedly along the length of the deposit, from $2.62 \times 10^{-5}$ to $0.63 \times 10^{-5} mm^3/Nm$, while the corresponding microhardness increased from 202.3 to 210.9HV. The yield and ultimate tensile strengths hardly varied along the deposition direction of AISI304.

During wear-testing, grain refinement occurred directly beneath the wear path. No appreciable differences in the yield strength occurred along various directions in the case of mild steel. In similar work[104], 3-dimensional structures of ER70S-6 mild steel wire were built up by depositing layers onto a 3mm-thick substrate of S235JR steel sheet. On the basis of hardness profiles and microstructural observations, it was concluded that the most favorable process conditions were those provided by cold-metal transfer. It was also noted that a continuous process was better than single cold-metal transfer.

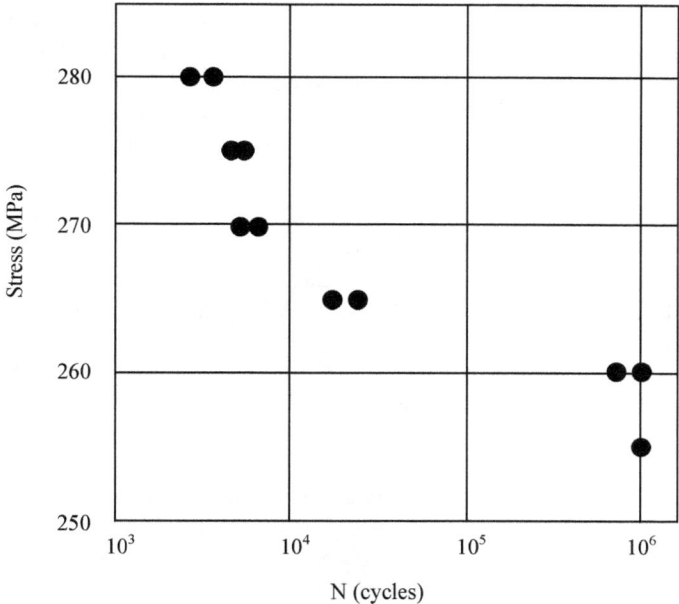

*Figure 16. S-N fatigue results for laser melt deposited SS304 stainless steel*

Laser engineered net shaping has been used[105] to manufacture compositionally gradated structures made from vanadium carbide and AISI304 stainless steel, with the carbide content ranging from 5 to 100wt%. Internal stresses, arising from the carbide content, markedly increased the hardness and wear-resistance of the coatings. The 100%VC outer layer increased the hardness by 1450HV and lowered the wear rate by 95%, as compared with the AISI304 substrate.

Walls of AISI304L, created using wire arc method, were subjected[106] to fatigue testing. Crack growth in the high-cycle regime was studied using horizontally- and vertically-oriented single-edge notch bend specimens taken from various positions in the wall. The R-ratio was 0.1 and the test frequency was 10Hz. A Paris-Law behavior which was similar to that of wrought steel alloys was observed in the vertical orientations exhibiting the greatest crack-growth resistance.

Selective laser sintering has been used[107] to create composites by sputter-coating tungsten carbide particles with a nanocrystalline thin film of stainless-steel as a metallic binder. Carbide powder was sputter-coated with 13wt% of AISI304L and subjected to various laser powers and scanning speeds. During processing, some of the carbon from the carbide diffused into the coating and some elements from the coating also diffused into the carbide particles, thus leading to the formation of phases such as $(Fe_{2.5}Ni_{0.3})(Cr_{0.7}W_{2.5})$.

The metal inert-gas wire-filling method was used[108] to deposit 304 stainless steel layer-by-layer onto a 20mm-thick low-carbon steel surface, using a voltage of 27V, a current of 190A, a deposition speed of 8.69mm/s, a wire-feeding speed of 7m/min, a wire-feeding angle of 90° and a shielding-gas flow-rate of 15l/min. With gradual reduction of the heat input, the grains in a given area were gradually refined. The cladding layer comprised columnar crystals, dendrites and equiaxed crystals and the microhardness increased as the heat input decreased, giving a maximum value of 257.8HV.

*AISI316*

A study of the effect of the laser energy-density during the selective laser melt processing of 316L stainless steel showed[109] that the surface roughness was affected mainly by the point distance; with an increased point distance leading to increased roughness: from 10 to 16μm. The hardness attained a maximum of 225HV at 125J/mm3, and was related to the porosity. That is, an increased porosity led to a decreased hardness. The energy density affected the total porosity, with the minimum porosity, 0.38%, corresponding to an energy density of $104.52J/mm^3$. Balling (particle coalescence) varied from small-ball features at low laser-energy densities to a mixture of small and large ball features at high energy densities.

Laser metal deposition, using a filler, is an effective route to additive manufacturing due to its high deposition efficiency, improved surface quality and lower loss of material. Single beads, and samples having 10 layers, were deposited[110] onto a 316L stainless-steel surface using a 4kW continuous-wave fibre laser and an arc-welding machine. Layered samples having a large deposition-height and smooth side-surfaces could be obtained by choosing suitable parameters. The microstructures exhibited fine austenite grains, with good metallurgical bonding existing between the layers, indicating an austenite solidification mode. Ferrite which was precipitated at the grain boundaries had a sub-grain structure with a fine uniform grain-size. The microhardness was greater, 205 to 226HV, in the middle of the deposition area. The tensile strength of a 50-layer sample attained 669MPa, and dimples at the fracture surface revealed the occurrence of ductile fracture.

Laser engineered net shaping has been used[111] to create compositionally-graded magnetic/non-magnetic bimetallic structures. This was done by combining non-magnetic AIAS316 austenitic stainless steel with magnetic AIAS430 ferritic stainless steel into a single structure, by using a high-powered laser to melt and bond metallic powder, in successive layers, to create the 3-dimensional structure. The microstructures had a preferred grain-growth direction at the interfaces of the deposited layers. The hardness varied smoothly, across the cross-section, from the highest value of 266HV in the AIAS430 region to the lowest value of 174HV in the AIAS316 substrate.

Wire-based directed energy deposition techniques permit the rapid production of large-scale structural components, which is not currently possible using powder-bed fusion methods. On the other hand, the larger melt-pool widths and higher energy inputs can lead to greater porosity, and other problems. The wire arc method was used[112] to produce austenitic stainless-steel single-bead walls which exhibited a marked wire-texture in the build direction, and anisotropic elastic moduli. The walls were of high quality, with a porosity of less than 0.32% and yield and tensile strengths of about 320.4 and 531.6MPa, respectively.

A laser/wire direct metal deposition system has been used[113] to prepare test-pieces of 316LSi which were of 2 types: thin-walled and block. The different thermal histories, due to different inter-layer time-intervals, had a marked effect upon the mechanical properties. Thus the thin-walled samples, with their lower cooling-rates, comprised coarser columnar grains and exhibited lower ultimate tensile strengths and lower hardness values as compared with those of block samples. By observing the melt-pool *in situ*, an empirical correlation could be established between the melt-pool area and the cooling rate. This then permitted feed-back control of the scale of the final solidification structure by maintaining a desired melt-pool size. Tensile samples were tested in orientations which were parallel to, or perpendicular to, the deposition direction and this showed that samples which were oriented in the perpendicular direction had lower ultimate strengths and elongations. This was attributed to a weaker bonding, at the interlayer/bead interface, due to lack-of-fusion pores.

The tensile properties of laser powder-bed-fused AISI316L stainless steel samples with a density greater than 98.8%, a large fraction (40-60%) of low-angle (2 to 10°) grain boundaries and an average grain size of 30 to 50µm, were investigated[114]. The yield strength ranged from 552 to 635MPa and the elongation-to-failure was 0.09 to 0.42 for as-processed samples. The corresponding values were 592 to 690MPa and 0.29 to 0.50 for samples machined from as-processed thin plates. A large scatter in the elongation of some samples was attributed to a sensitivity of the thin test-samples to the presence of flaws. There was a markedly higher strain-rate sensitivity (0.02 to 0.03) of powder-bed fused material as compared with that (0.006) of coarse-grained samples. The activation volume was only 20 to 30b$^3$. The tensile plasticity of the additively manufactured steel appeared to be controlled by a finer length-scale than the observed grain size.

The ability of selective laser melt processing to form multi-material bi-metallic structures was demonstrated[115] by creating 316L/CuSn10 structures. The width of the fusion zone was about 550μm, and dendritic crack sources were found at the boundary between the bi-metallic fusion zone and the steel. The Vickers microhardness in the direction perpendicular to the interface changed gradually, from 233.1HV in the steel zone, to 154.7HV in the bronze zone. The ultimate strength of the 316L/CuSn10 joint was 423.3MPa, as compared with the 673.1MPa of the stainless steel and the 578.7MPa of the bronze. The fusion zone between steel and bronze exhibited brittle fracture. Three-point bending tests were used to determine the interfacial bond strength, showing that the maximum flexural strength of the 316L/CuSn10 bi-metallic structure was not between, but below, that of 316L and CuSn10.

The size of the melt-pool, and the temperature distribution around it, are critical factors determining process characteristics such as the deposition rate. A parametric study was made[116] of the effects of laser scanning-pattern, power, speed and build-direction during powder-bed fusion additive manufacturing upon residual stresses in the final product. This involved destructive surface residual stress measurements, combined with a non-destructive volumetric evaluation method. Good agreement was found between the results deduced using the two measurement techniques. A reduction in the residual stress was obtained by decreasing the scan island size, increasing the island-to-wall rotation to 45 and increasing the applied energy per unit length. Neutron diffraction data revealed that in-plane residual stresses were affected by scan island rotation, but axial residual stresses were unchanged. The in-plane behavior was attributed to misalignment between the maximum thermal stress, which occurred in the scan direction, and the largest dimension of the component.

In selective laser melting, due to the thermal gradients involved, the products are exposed to residual stresses that then lead to distortion. The process was used[117] to prepare 316L stainless-steel compacts, which were then subjected to heat treatments at 650, 950 or 1100C for 2h. The microhardness of the as-prepared material was between 209.0 and 212.2HV, and thus much higher than that of the heat-treated material.

Laser engineered net shaping methods have been used[118] to form single-bead multilayer AISI316L stainless steel structures by using a given laser power and scanning speed. Differing deposition efficiencies resulted from adjusting the powder flow rate and layer incrementation. For certain laser-powers and scanning-speeds, the deposition efficiency increased from 12.41 to 22.62mm$^3$/s with increasing powder flow-rate and layer incrementation. The laser energy which was expended decreased from 98.84 to 53.06J/mm$^3$ and the energy efficiency increased by 46.32%. The microstructures comprised columnar dendrites, with the dendrite length increasing with deposition efficiency. The properties, for various deposition efficiencies, were consistent and did not decrease with deposition efficiency, the tensile strength and yield strength for instance being stable at 510 and 290MPa, respectively. The elongation was about 40%, and the microhardness was about 180HV.

Selective laser melting was used[119] to prepare AISI316L stainless-steel composites, containing 0, 5, 10 or 15wt% of calcium silicate, for biomedical purposes. The addition of CaSiO$_3$ particles markedly affected the microstructure and properties of the specimens. Their tensile strengths ranged from 320 to 722MPa, while the microhardness and elastic modulus were 250HV and 215GPa, respectively. The composites were ductile, and the fracture mode was a mixture of ductile and brittle fracture.

Rapid scanning rates were used[120] to produce high-density AISI316L stainless steel by means of selective laser melting, with the object of increasing the production-rate while maintaining low porosity. Densities of better than 99% were obtained in every case, and the build-rate was improved by about 72% as compared with the usual processing methods. The microhardness of the resultant parts was between 213 and 220HV, and this was much higher than the 155HV which was usual for standard preparation methods.

In a typical case, selective laser melting of 316L, using a laser power of 25 to 225W, scanning speed of 50 to 320mm/s and scan spacing of 0.04 or 0.06mm, produced material having a density of 97.6% and a hardness of 220HV[121].

In an early study, direct metal laser deposition was used[122] to create near net-shape austenitic stainless steel structures which possessed a hardness of 397HV. This was satisfyingly high when compared with the 223HV which was typical of commercial AISI316 plate. The present austenitic structure had a reduced modulus of 126GPa and a hardness of 4.95GPa[123]. Cyclic polarization corrosion-test results showed that the pit-growth resistance of both materials was similar.

The metal powder's flowability can also have a critical effect upon the densification, and therefore upon the mechanical properties, of the product. Modification of the powder surface by the presence of a nanocrystalline thin film, for example, can cause a marked improvement in the flowability. Gas atomisation is a popular method for manufacturing free-flowing metal powders for many of the additive manufacturing processes. Very fine (<20pm) gas-atomised powders can be used in high-precision 3-dimensional selective laser melting and this leads to excellent component precision and good surface finish. A new binder jetting process for creating metal or ceramic parts was developed[124] by using coated-powder and ink-jet techniques which involved ceramic particles coated with 100nm of water-soluble resin and an ink which contained a cross-linking agent that acted on the coated resin but did not have any binding ingredients. It was possible to produce 316L stainless steel parts by using this process. Lithography-based additive manufacturing processes for metals can similarly be based upon using a photo-reactive metal suspension which is cured by selective exposure to light. The photo-reactive suspension is loaded with up to 50vol% of commercially available 316L powder and, after printing the green material, heat treatment is used to impart the final metallic properties.

Solid specimens having a relative density of 99.8% were prepared[125] from gas-atomized powder. Following electron-beam processing, the phase remained austenitic and the composition remained essentially unchanged. Tensile and Charpy-V tests, performed at 22C and 250C, showed that the yield stress, ductility and toughness were well above the criteria required for nuclear applications; apart from the ultimate tensile strength. The microstructure contained solidified melt-pools, columnar grains and irregularly-shaped sub-grains. Copious precipitates, enriched in chromium and molybdenum, were present at columnar grain boundaries but no element segregation was found at the sub-grain boundaries. This microstructure reflected the occurrence of non-equilibrium rapid solidification and of annealing during the anisotropic flow of heat within the powder matrix. Defects were created due mainly due to a large (100μm) layer thickness which led to insufficient bonding between adjacent melt-pools during processing.

A study of 316L prototyping, using 3-dimensional laser engineered net shaping, concluded[126] that the final behaviour was dictated by interactive metallurgical reactions during powder feeding, molten metal flow and solidification. The process is capable of creating high-strength ductile stainless-steel components due to the fine cellular spacing which results from rapid solidification cooling, and to the well-fused epitaxial interfaces at interpass boundaries. On the other hand, without further optimization the deposits could suffer localized hardness variations arising from an heterogeneous microstructure. In particular, deposits could exhibit lower strain and premature interpass delamination parallel to the build direction, due to the existence of an interpass heat-affected zone, feedstock inclusions and porosity arising from incomplete molten metal fusion.

Additively manufactured metals can be highly textured and contain large columnar grains which nucleate epitaxially in high temperature gradients and rapid solidification situations. Such microstructures account for some of the large observed differences between the properties of additively manufactured and traditionally prepared alloys, where equiaxed grains would be preferred. In a study[127] of the effect of the laser intensity profile on 316L stainless steel it was generally found that columnar grains formed with increasing laser power and scan-speed, regardless of the beam profile. When conduction-mode laser-heating occurred however, the scanning parameters which resulted in coarse columnar microstructures when using Gaussian profiles could produce equiaxed or mixed equiaxed-columnar microstructures when using elliptical profiles. By using *ad hoc* spatial laser-intensity profiles, localized microstructures and properties could be introduced into additively manufactured parts.

Austenitic stainless steel samples were produced[128] by using the gas metal arc additive manufacturing method, revealing that a large number of well-aligned austenitic dendrites were oriented vertically, forming large columnar grains in the middle, while some dendrites were bent towards the plate surfaces and formed small columnar grains near to the edges. The microstructures consisted of δ-, γ- and σ-phases. After one layer had been deposited, the δ-phase had a reticular morphology within austenitic dendrites. The δ-phase re-dissolved in austenite, with intermetallic σ-phase forming at γ/δ interfaces over 3 layers during thermal cycling. Under thermal influences, after the fourth layer, the δ- and σ-phases turned into fine vermicular morphologies within austenitic dendrites. The tensile properties of the additively manufactured steel were comparable to those of wrought steel. Microcracks nucleated in the interior of σ-phases and grew into large cracks. The fracture mode was ductile, as indicated by obvious fracture-surface dimples. The mechanical and corrosion properties of gas metal arc additively manufactured AISI316L could be optimized[129] by adjusting the volume fractions of σ- and δ-phases by means of heat treatment. Treatment at 1000 1200C for one hour did not obviously affect the grain morphology but did affect the contents of σ- and δ-phase. Treatment at 1000C increased the amount of σ-phase in steel, causing both the ultimate tensile strength and yield stress to increase while decreasing the elongation and reduction-of-area. Treatment at 1100 to 1200C completely eliminated the σ-phase, leading to a decrease in the ultimate tensile strength and yield stress but an increase in the elongation and reduction-of-area. The σ-phase exerted a better strengthening effect than did the δ-phase, but could impair the ductility and increase the possibility of crack generation. Limiting both the σ and δ phases by heat treatment could improve the corrosion resistance, although the σ-phase seemed to have a detrimental effect upon the corrosion resistance than did the δ-phase.

In the case of selective laser-melt processed pure stainless steel, directional columnar grains were found. The addition of $TiB_2$ nanoparticles to the steel matrix greatly reduced the size of the melt-pool and of the grains however, and spoiled the directionality[130]. There was no compositional difference between the boundary and interior of the melt-pool areas of processed $TiB_2$/316L nanocomposites; thus suggesting that the elements did not segregate macroscopically. At higher $TiB_2$ contents, the alloying elements microsegregated at the boundaries of cellular structures due to particle accumulation. There were mainly cube-like $TiB_2$ nanoparticles at the boundaries and in the interior of cellular structures. Except at 600C, where embrittlement occurred, the nanocomposites had a high compressive yield stress and ductility at room and elevated temperatures. Conventional strengthening effects could not explain all of the yield stress increase.

In the powder-based laser directed-energy deposition method, a laser beam melts a millimetre-scale metal pool into which powder is sprayed. The powder is first held by surface tension, and floats briefly before melting. A simple equation[131] gives a rapid prediction of the time for which the powder remains on the surface. The sensitivity of the processing method to powder-scale surface wettability explains why control of the feedstock powder properties is required for ensuring reliable product quality.

In a new additive manufacturing process for making austenitic stainless steel parts[132], they are built up by using an extrusion-based three-dimensional printer loaded with a metal–polymer composite filament, before applying standard sintering processes.

As an example of a practical application, powder-bed laser fusion additive manufacturing has been considered[133] for producing stainless steel heat pipes and thus improving heat transfer in a hydrogen-isotope separation unit where the temperature varies cyclically from cryogenic to moderate. This could produce consistent components having acceptable tensile properties, but full density was not always achieved and minor modification was required in order to correct this. A previous challenging application had been the creation of an air vent which was hollow, thin-walled, of variable cross-section and of limited size. It was made from stainless steel by using a 4kW laser powder-fed additive manufacturing system[134].

A further possibility is that of using electron-beam powder-bed fusion additive methods to apply face-centered cubic nickel-based Inconel-690 cladding to face-centred cubic 316L stainless-steel so as to impart improved mechanical properties and a reduced sensitivity to corrosion[135]. The constitutional supercooling solidification phenomena associated with the process promoted the appearance of [100]-textured columnar grains with their lower-energy boundaries rather than random high-angle grain boundaries, but no coherent {111} twin boundaries. Irregular low-energy sub-grains (2 to 3μm) were formed, together with dislocation densities ranging from $10^8$ to $10^9$/cm$^2$ and an homogeneous distribution of $Cr_{23}C_6$ precipitates. The latter formed within grains but not in sub-grain or columnar grain boundaries. Such microstructures imparted a tensile yield stress of 0.527GPa, an elongation of 21% and a Vickers microhardness of 2.33GPa to the Inconel cladding, as compared with the equivalent figures of 0.327GPa, 53% and 1.78GPa for the wrought 316L substrate. Aging (685C, 50h) of the cladding and the substrate precipitated $Cr_{23}C_6$ carbides in the Inconel columnar grain boundaries, but not in the low-angle low-energy sub-grain boundaries. The $Cr_{23}C_6$ carbides precipitated in stainless-steel grain boundaries but not in low-energy coherent {111} twin boundaries.

*Table 17. Comparison of the mechanical properties of selective laser melted and conventionally prepared grade 300 maraging steel*

| Orientation | Condition | UTS (MPa) | YS (MPa) | Elongation (%) | HRC |
|---|---|---|---|---|---|
| horizontal | as-processed | 1165 | 915 | 12.4 | 34.8 |
| horizontal | aged | 2014 | 1967 | 3.3 | 54.6 |
| horizontal | solutionized | 1025 | 962 | 14.4 | 29.8 |
| horizontal | solutionized&aged | 1943 | 1882 | 5.6 | 53.5 |
| vertical | as-processed | 1085 | 920 | 11.3 | 35.7 |
| vertical | aged | 1942 | 1867 | 2.8 | 52.9 |
| vertical | solutionized | 983 | 923 | 13.7 | 27.5 |
| vertical | solutionized&aged | 1898 | 1818 | 4.8 | 51.3 |
| | wrought | 1000–1170 | 760–895 | 6–15 | 35 |
| | wrought&aged | 1930–2050 | 1862–2000 | 5–7 | 52 |

Selective laser melting was used[136] to test the effect of the powder feedstock upon the properties of 316L products. Gas-atomized powders having 3 different particle-size distributions were tried. Microstructural examination revealed structures consisting of solidified melt-pools, columnar grains and multiform sub-grains. Products which were made using fine powder exhibited the best mechanical properties, with an ultimate tensile strength of 611.9MPa and a yield stress of 519.1MPa, together with an elongation of 14.6% and a microhardness of $291HV_{0.1}$.

A study of selective laser melt-prepared 18Ni-300 maraging steel showed[137] that aging had a marked effect upon the strength and wear-resistance. The optimum aging conditions which were required in order to maximise the strength and wear-resistance were a temperature of 490C and a time of 3h. A lower or higher aging temperature led to under-aging or over-aging effects, thus reducing the strength and wear-resistance. Shorter or longer aging-times also led to a decrease in strength and wear-resistance. The variations in the mechanical and friction properties were due mainly to changes in composition and microstructure. Cold-spray and selective laser melting techniques have been used[138] to produce maraging steel 300 composites which were reinforced with WC particles. The selective laser melted composite had a lower porosity than that of the cold-spray composite, but the latter exhibited a slightly greater microhardness. The selective laser-melted composite had a markedly lower wear-rate than that of the other composite.

Nearly fully-dense selective laser melted grade-300 maraging steel[139], subjected to various heat treatments, had an average grain size of $0.31\mu m$; thus suggesting a cooling-rate of up to $10^7 K/s$. Massive needle-shaped nanoprecipitates of $Ni_3X$, where X was titanium, aluminium or molybdenum, were prominent in the martensitic matrix and accounted for the observed age-hardening. The strengthening was attributed to the Orowan bowing mechanism and coherency strain hardening. A build-orientation anisotropy due to the layering was observed in as-processed and heat-treated material (table 17). Heat treatment led to strengthening and also markedly relieved residual stresses; thereby partially reducing the anisotropy. It also affected the tribological properties (table 18)

*Table 18. Tribological properties of selective laser melted grade 300 maraging steel*

| Condition | Coefficient-of-Friction | Wear-Rate ($mm^3/Nm$) |
|---|---|---|
| as-processed | 0.62 | $4.43 \times 10^{-8}$ |
| aged | 0.56 | $1.91 \times 10^{-8}$ |
| Solutionized&aged | 0.58 | $2.53 \times 10^{-8}$ |

Direct metal laser sintering of metal powders was used[140] to produce specimens of EOS maraging steel, some specimens being subjected to age-hardening (490C, 6h, air cooling). Heat-treated and as-processed specimens were taken at 0° and 90° with respect to the specimen axis. Static tensile tests indicated that there was no difference in the mechanical properties of 0° and 90° specimens, whether as-processed or heat-treated (table 19). Axial fatigue data indicated that the lowest fatigue strength occurred in 0°-oriented specimens. The fatigue strength of as-processed 0°-oriented specimens was slightly lower than that of heat-treated specimens having the same orientation. In the case of 90°-oriented specimens, the scatter of the results did not allow differences to be clearly distinguished. Upon taking the fatigue strength at $5.0 \cdot x \ 10^5$ cycles to be a reference point, the untreated 0° and 90° specimens had 72 and 33% lower fatigue strengths, respectively, as compared with those of the vacuum-melted maraging steel in the annealed condition. The heat-treated 0° and 90° specimens had 68 and 61% lower fatigue strengths, respectively, when compared with those of vacuum-melted maraging steel in the annealed and age-hardened condition.

*Table 19. Properties of direct metal laser sintered EOS300 maraging steel*

| Condition | Orientation (°) | E (GPa) | UTS (MPa) | 0.2%YS (MPa) | Elongation (%) |
|-----------|-----------------|---------|-----------|--------------|----------------|
| as-processed | 0 | 169.05 | 1213 | 1120 | 33.6 |
| heat-treated | 0 | 187.01 | 2029 | 1950 | 10.3 |
| as-processed | 90 | 170.62 | 1215 | 1030 | 29.1 |
| heat-treated | 90 | 187.46 | 2055 | 1950 | 14.3 |

When selective laser melting was used[141] to process a maraging steel (Fe-18Ni-9Co-5wt%Mo), it was found that there existed a relatively large processing-window within which products possessing a relatively high relative density and good surface quality could be obtained. The as-prepared specimens comprised a martensite matrix, with trace amount of austenite. The amount of austenite increased during aging because of reversion of the martensite to austenite. Solution and aging treatments led to the elimination of the austenite and to the formation of intermetallic precipitates in the martensite matrix. As-prepared and aged specimens had almost the same average grain size, but solution and aging treatments led to grain growth of the martensite matrix and to a marked change in grain orientation. Specimens with their build direction parallel to the tensile loading direction exhibited much lower elongations than did those for which the build direction was perpendicular to the loading direction. The maximum tensile strength and hardness were 2033MPa and 618HV, respectively, following solution treatment (820C, 1h) and aging (460C, 5h).

Due to the nature of the layer-by-layer build-up of additively manufactured parts, the already-deposited layers experience a cyclic re-heating in the form of sequential temperature-pulses. This effective heat treatment resource can be exploited[142] by inducing the precipitation of NiAl nanoparticles in Fe-19Ni-x(at%)Al maraging-type steel. Laser metal deposition was used to create compositionally graded specimens having aluminium contents ranging from 0 to 25at%. In order to guarantee existence of the desired martensitic matrix, an upper bound of 15at% was put on the aluminium concentration. Due to the presence of NiAl precipitates, a lower bound of 3 to 5at%Al was established. Within this concentration window, increasing the aluminium concentration increased the hardness by 225HV due to the very high number-density of $10^{25}$ NiAl precipitates/m$^3$.

*AISI H11*

Selective laser melting has been used[143] to prepare specimens of the H11 hot-working tool steel, and of a leaner version, L-H11, of the alloy. The rapidly solidified microstructures responded to precipitation-hardening annealing without requiring any solution-annealing treatment. The microstructures of the as-prepared alloys consisted of α-Fe dendritic cells with carbon-rich γ-iron regions at the grain boundaries. Air-quenching transformed the solidification cells into lath-martensitic structures and formed an $M_3C$ phase which further changed into more complex carbides upon tempering. The hardness of the quenched and tempered H11 steel was similar to that obtained by processing the alloy conventionally, and the final hardness differential between the H11 and L-H11 alloys was limited to 62HV. Laser metal deposition was used[144] to process AISI H11 (1.2343, X37CrMoV5-1) tool steel powder which contained carbon-black nanoparticles. This showed that adding carbon-black nanoparticles increased the incidence of martensitic and bainitic phases, as well as the precipitation of carbides at grain boundaries. This led in turn to a marked increase in the hardness and compression yield stress, of the order of 11% over that of samples made from unmodified powder. It was moreover feasible to produce parts having a variable hardness from layer-to-layer. The reduced-energy gas metal arc welding process, a cold-metal transfer method, has been used[145] to prepare specimens of the hot-working tool steel, X37CrMoV5-1. This showed that the steel could be used to generate crack-free 3-dimensional components, with high reproducibility and near-net shape, at a deposition rate of up to 3.6kg/h. The mechanical properties were determined the thermal field, which was in turn controlled by the bypass temperature and the electric arc energy. One factor which was critical to the welding process was the energy per unit length. When the bypass temperature was above the martensite-start temperature, there existed an homogeneous hardness level along the structure provided that the energy generated by the welding arc was sufficient to keep the temperature of all of the layers above the martensite-start temperature.

*AISI H13*

kLaser additively manufactured AISI H13 tool steel was studied[146] in the as-prepared state and after annealing at 350, 450, 550, 600 or 650C for 2h. The microstructure of the as-deposited material consisted of martensite, ultrafine carbides and retained austenite. During the heat-treatment, the martensite was converted into tempered martensite and fine alloy carbides which precipitated. Following treatment at 550C, the greatest hardness and nanohardness were $600HV_{0.3}$ and 6119.4MPa (table 20), due to the many needle-like carbide precipitates. The hardness first increased firstly, and then decreased, with increasing annealing temperature. When the temperature exceeded 550C, the carbides coarsened and the martensitic matrix recrystallized at 650C. The least impact energy was 6.0J at 550C. Samples which were treated at 550C exhibited a greater wear volume-loss than did the others.

*Table 20. Properties of H13 tool steel in various conditions*

| Condition | Young's Modulus (GPa) | Hardness (MPa) |
|---|---|---|
| as-prepared | 5578.5 | 317.87 |
| annealed (350C, 2h) | 5679.0 | 318.25 |
| annealed (450C, 2h) | 5856.6 | 312.44 |
| annealed (550C, 2h) | 6119.4 | 314.76 |
| annealed (600C, 2h) | 5428.9 | 313.54 |
| annealed (650C, 2h) | 5145.4 | 326.56 |

A 3-dimensional heat-transfer and fluid-flow model of the wire arc process has been used[147] to calculate temperature and velocity fields, deposit size, and other factors. The calculated fusion-zone geometries and cooling rates, as a function of arc power and travel speed agreed well with experimental data on H13 tool-steel deposits. It was deduced that convection is the main heat-transfer mechanism within the melt pool. More rapid travel increased the cooling rate but reduced the ratio of temperature gradient to solidification rate, thus leading to solid/liquid interface instability. Higher deposition rates were preferably achieved by increasing the heat input, using thicker wires and increasing the wire feed-rate. When using limited laser power, laser re-melting increases the relative density and hardness of H13 tool steel. Single melt-pool analysis shows that pores are located in the upper region of the melt pool when the scanning speed greater than 400mm/s. A low scanning speed, such as 200mm/s, generates keyhole-type pores beneath the melt pool when there is a high energy-input. During secondary laser scanning, pores in the upper region of the melt pool can be efficiently closed by using a suitable scan speed. Those pores beneath the melt pool cannot be removed by secondary scanning. When each layer of a 3-dimensional construction is re-melted, the relative density and hardness improve under most conditions. Optimum settings[148] led to a maximum relative density of 99.94% and a Rockwell-C hardness of 53.5. Laser-aided direct metal deposition has been used to repair[149] AISI H13 samples containing hemispherical defects. The repair was effectuated by using cobalt-based alloy powder as a filler. Microstructural observations and tensile tests confirmed that a strong metallurgical bond existed at the interface. A defect-free columnar structure predominated in deposits close to the interface, while other parts of the deposit consisted of a dendrite structure with interdendritic eutectic. The repaired samples had a higher ultimate tensile strength and lower ductility as compared with the base metal. The repaired samples fractured in a brittle manner in the deposit, and cracks propagated along the grain boundaries. The hardness of the deposited layers was much greater than that of the substrate. A study was made[150] of the effects of the microstructure and defect content upon the mechanical properties of 3-dimensionally printed H13 tool steel which had been prepared by means of selective laser melting using a scan-speed of 200mm/s and a layer thickness of 25µm. Small keyhole-like pores were observed, giving a porosity of 0.4%, where the lowest porosity was in the central region. The hardness was about 550HV and the yield and tensile strengths were 1400 and 1700MPa, respectively. A thermographic study has been made[151] of the effect of build-height upon the properties of H13 tool steel during selective laser melting. The material was pre-heated to 550C, and an infra-red camera recorded a temperature drop of some 170C along the height, which affected the hardness and microstructure. The former was between 614 and 662$HV_{10}$, while a longer period of time at higher temperatures in the lower portion of the component favoured the formation of martensitic structures. No effect upon the surface roughness, which was between 1.5 and 2.1µm, could be detected.

*17-4PH*

Direct metal laser sintering was used[152] to produce 17-4 stainless-steel components which were then shot-peened. The as-processed material contained a high content of retained austenite, and the shot-peening treatment provoked a martensitic phase transformation. Micro-strains and compressive residual stresses were markedly increased by the shot-peening. As compared with as-processed material, shot-peened samples had a refined surface microstructure and this imparted a more favorable roughness, hardness, compressive yield stress and wear resistance. Pulsed metal inert gas welding, with a tandem torch, was used[153] to deposit the 17-4PH martensitic stainless steel. The mechanical and metallurgical properties were explored in order to identify the maximum deposition-rate which did not produce defects. This limit was placed at 9.5kg/h. Material extrusion additive manufacturing can produce metal products when combined with sintering. A polymer filament loaded with 55vol% of 17-4PH stainless steel powder has been used to print dog-bone specimens which were then sintered to give metallic specimens[154]. The average Young's modulus was 196GPa, the average maximum stress was 696MPa and the elongation-to-fracture was 4%.

Plates of AF1410 ultra-high strength steel were made[155] by means of laser additive manufacturing. The microstructure of as-deposited samples exhibited features of directional solidification, with an associated hardness of about 360HV. During heat treatment, the directionally solidified microstructure disappeared, leaving a refined microstructure and tempered martensite; imparting a hardness of about 510HV. The room-temperature yield stress and tensile strength of the heat-treated steel attained 1490 and 1610MPa, respectively, together with an elongation of 12.8% and a reduction in area of 70%.

Pulsed-current shaped metal deposition was used[156] to produce components from AISI308LSi stainless-steel wire. The structure of pulsed-arc current produced specimens was generally finer-grained, with a high residual ferrite content and an absence of columnar grains. Good metallurgical bonding, with no sensitization effects, was found.

Magnetic alloys with the composition, $(Fe_{60}Co_{35}Ni_5)_{73.5}Si_{13.5}B_9CuMo_3$, known for a good saturation magnetization and mechanical properties have been prepared[157] by means of laser additive manufacturing. Laser re-melting had a beneficial effect and, following laser re-melting, the alloy had the higher saturation magnetization of 165.5emu/g and the lower coercivity of 12Oe. The microstructure of the re-melted alloy comprised mainly dense and homogeneous dendrites and was predominantly body-centered cubic. As a result of a reduced grain size, the wear resistance and microhardness had greatly increased.

The maximum thickness of bulk metallic glass components which can be created using traditional processes is limited, whereas laser-based powder-bed additive manufacturing methods can potentially produce relatively large iron-based bulk metallic glass specimens. For example, additively manufactured specimens can exceed the critical casting thickness by a factor of more than 15, and it seems[158] that a fully amorphous structure can be maintained throughout.

Selective laser melted high strength steel, AISI4130, was used[159] to construct micro-lattices. The bulk steel had a yield strength of 1243MPa, an ultimate tensile strength of 1449MPa and an elongation-to-fracture of 15.5%. The microstructure consisted of an alternating tempered and well-retained martensitic network. This unique microstructure resulted from applying a single-step fusion and quenching process, followed by an *in situ* rapid dynamic tempering effect which was linked to the laser-scanning patterns. Orthogonally isotropic micro-lattices were designed and optimized using finite-element calculations. Their energy-absorption per weight and volume ranged from 13 to 35J/g and from 12 to 76J/cm$^3$, respectively, for relative densities of 10 to 30%. The energy absorption efficiency was about 80%.

Composite specimens of TiN-reinforced AISI420 stainless steel were prepared[160] using selective laser melting. The addition of TiN had a marked effect upon the density. At higher laser powers, the diffusional behavior of the TiN became more marked and densification of composites was increased to a maximum relative density of 98.2%. The best composites had a Rockwell-C hardness of 56.7; 11.8% higher than that of the plain steel when produced in the same way.

The application of the so-called TOPTIG wire arc additive manufacturing to ER2594 super duplex stainless steel (Fe-26Cr-10Ni-3wt%Mo) has recently been explored by creating a wall component[161]. A relatively high (98.45wt%) austenite content existed in the wall-body, but little (1.55wt%) ferrite. A massive growth of austenite in the wall-body region was due mainly to a continuous propagation, of intergranular secondary austenite, which resulted in coarsening of the primary austenite. The growth of intergranular secondary austenite and the precipitation of intragranular secondary austenite were due to the precipitation of chromium nitrides. Deleterious phases such as $\lambda$ and $\sigma$ did not greatly affect the fracture behaviour of as-deposited material. On the other hand, the precipitation of CrN and the presence of $Cr_2N$ caused an uneven distribution of elongation and impact-toughness values throughout the sample. The 0.2% yield stress was 530MPa, the ultimate tensile strength was 852MPa and the elongation was 35%. The impact toughness ranged from 104.21 to 122.24J/cm$^2$. Variations in the properties were attributed to the differing fractions of austenite and ferrite. Nitrogen tended to decrease the anisotropy of the yield stress, due to nitrogen work-hardening, but the nitrogen had no discernible effect upon the ultimate tensile strength or the elongation.

Samples of M2 (Fe-W-Mo-Cr-V) steel were produced[162], using laser additive manufacturing, and tempered for various times at 560C. The microstructures comprised fine equiaxed grains, dendrites and an interdendritic network of eutectic carbides, together with supersaturated martensite, retained austenite and $M_2C$-type carbides. The fraction of retained austenite gradually decreased with increasing tempering time. The microhardness of as-deposited samples was 688HV, while the hardness following a first, second and third tempering period was 833, 710 and 740HV, respectively. The samples exhibited an adhesive wear mechanism, and non-tempered material had a friction coefficient of 0.52. Following double tempering (560C, 2h), samples exhibited a larger wear volume-loss than did the others.

Laser melt deposited crack-free parts were additively manufactured using powder made from hot-working tool steel, 1.2344 (Fe-5Cr-1,5Mo-1wt%V). The products[163] had a dendritic solidification structure containing grain-boundary carbides. The hardness ranged from 400 to 750HV.

*Table 21. Properties of wire arc cold metal transfer prepared Fe-23Cr-18Mn-2wt%Ni*

| Condition | Orientation | YS (MPa) | UTS (MPa) | Elongation (%) |
|---|---|---|---|---|
| as-prepared | horizontal | 575.4 | 860.6 | 34.9 |
| as-prepared | vertical | 569.5 | 851 | 46.8 |
| heat-treated: 1050C, 300s | horizontal | 675.1 | 892.2 | 28.6 |
| heat-treated: 1050C, 300s | vertical | 679.8 | 931.6 | 39.5 |
| heat-treated: 1100C, 0.5h | horizontal | 442.2 | 663.5 | 16.5 |
| heat-treated: 1100C, 0.5h | vertical | 483.2 | 796.2 | 34.3 |

Cold metal transfer methods were used[164] make high nitrogen austenite stainless steel parts by using Fe-23Cr-18Mn-2wt%Ni wire with a nitrogen content of 0.99wt%. Excellent tensile properties were observed (table 21) for parts having a high nitrogen content. Amorphous inclusion islands and MnO microspherical inclusions were found. An increasing density of the latter, with diameters ranging from 0.1 to 1μm, impaired the tensile properties because the matrix/inclusion interfaces acted as preferred nucleation sites for $Cr_2N$ during heat treatment. Due to the stable austenite and the nitrogen work-hardening effect, planar dislocation-arrays predominated over the dislocation slip mechanism which, to some extent, diminished the strength anisotropy in various directions. Ferrite dendrites led to variations in the ultimate tensile strength and elongation by acting as conduits for cracks in the case of horizontal-direction samples.

A steel plate, 120mm long and 210mm high, has been manufactured[165] by means of wire arc additive manufacturing, using ER347 (Fe-20Cr-10wt%Ni) wire and gas metal arc welding. The plate was well-formed and no evident boundaries were observed between the layers. The hardness of the plate varied from top to bottom and was between 203.5 and 248.2HV. The microstructure comprised columnar dendrites and equiaxed dendrites within the multilayer deposits, and the volume fraction of delta ferrite in the as-deposited plate was 4.2%. Intermetallic compounds such as NbC were present. The properties were anisotropic, with a the sample taken at 45° having a higher tensile strength than those of 0° or 90° samples. Fractured tensile specimens in the as-deposited state exhibited dimple-like ductile-fracture surfaced.

Selective laser melted porous materials tend to exhibit an inferior fatigue behaviour, particularly with regard to the fatigue endurance ratio. It is found[166] that hot isostatic pressing (1000C, 150MPa) could eliminate pores but also decrease the microhardness from 403 to 324HV and the yield strength from 143 to 100MPa. The fatigue strength and endurance ratio at $10^6$ cycles were increased by hot isostatic pressing, with the fatigue endurance ratio at $10^6$ cycles changing from 0.3 to 0.55. This made it comparable to that of the solid material. The improvement was attributed mainly to a transformation from brittle $\alpha'$-martensite to tough $\alpha$ plus $\beta$ mixed phases due to the hot isostatic pressing. The tougher $\alpha$ plus $\beta$ mixture tended to resist the propagation of fatigue cracks by blunting them.

Selective laser melting has been used[167] to process high-strength Fe-8Mo-4Cr-2wt% tool steel, thus producing crack-free and highly dense (99.99%) parts. The effect of the localized high heat-input, applied over very short times, resulted in a fine homogeneous microstructure of martensite, austenite and carbides. This imparted a microhardness of $900HV_{0.1}$ and a compressive strength of about 3800MPa, combined with a fracture strain of up to 15%.

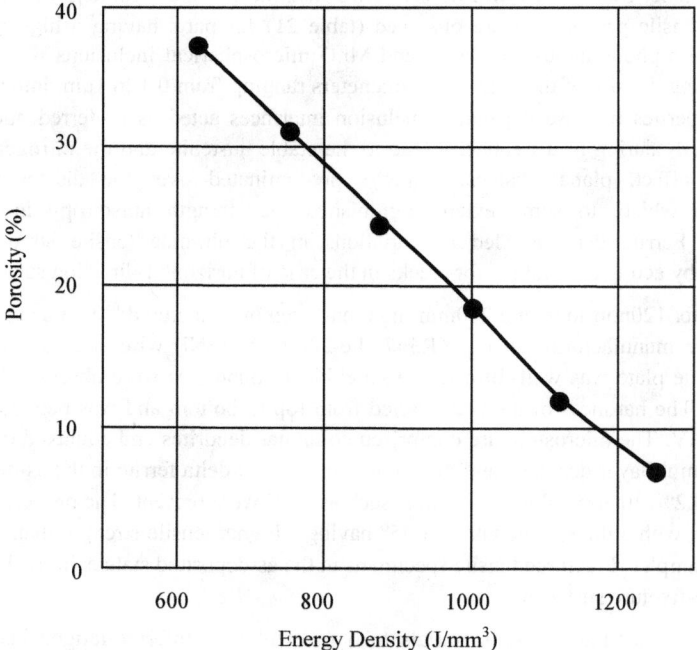

*Figure 17. Porosity of laser additively produced Mg-2wt%Ca samples as a function of the laser energy density*

The effects of process parameters, such as the inter-layer temperature, wire feed-rate, travel speed and ratio of wire feed-rate to travel speed, uon the surface roughness of thin-walled steel parts were investigated[168]. With other parameters fixed, decreasing the inter-layer temperature could be linked to an increase in the surface quality. With the ratio of wire feed-rate to travel-speed fixed, the surface roughness increased with increasing wire feed-rate. A lower feed-rate, matched with a lower travel speed, could decrease the surface roughness.

Three-dimensional plasma-metal deposition has been used[169] to combine the super duplex steel, 1.4410, with the austenitic steel, 1.4404. The properties of the transition zone lay between those of the two alloys, with the strength of the transition zone being higher than that of the austenitic material. It was concluded that the production of graded steel structures between 1.4404 and 1.4410 was feasible, and that mixing of the materials in the transition zone did not weaken the component.

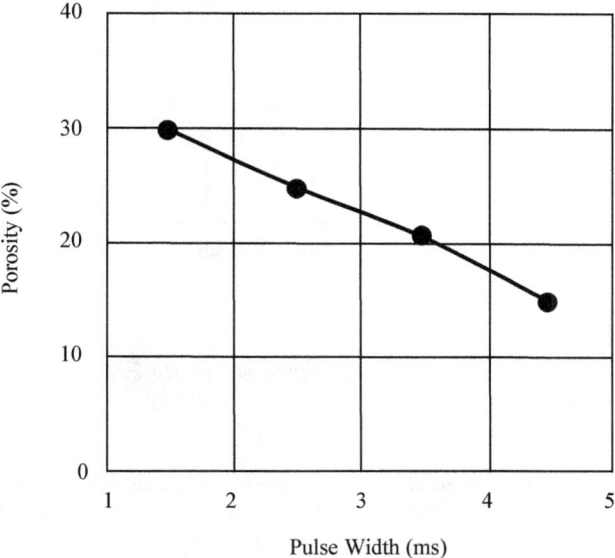

*Figure 18. Porosity of laser additively produced Mg-2wt%Ca samples as a function of the laser pulse-width*

## Magnesium

Additively manufactured components made from magnesium alloys are gaining in popularity as they are increasingly satisfactory replacements for more expensive titanium and cobalt alloys. Also significant is its applicability to implant-use in orthopaedic surgery as it is biodegradable, introduces very little stress and avoids the need for secondary surgery.

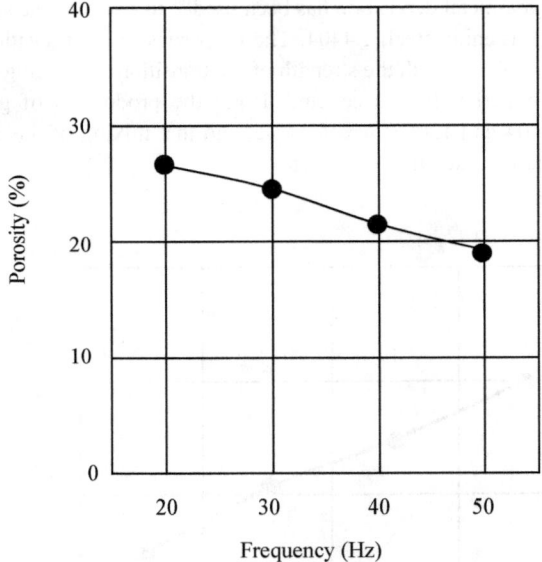

*Figure 19. Porosity of laser additively produced Mg-2wt%Ca*
*samples as a function of the laser frequency*

Laser additive manufacturing was used to produce porous Mg-2wt%Ca specimens, and this showed that the porosity and surface morphology depended upon the laser energy density (figure 17)[170] and other parameters (figures 18 and 19). When the latter was between 875 and 1000J/mm$^3$, the porosity of the alloy was between 18.48 and 24.60%. The product exhibited overlapping cladding lines and periodic morphological features. Further away from the melt pool, the grains underwent a transition from equiaxed to columnar and the grain size increased. All of the samples consisted only of α-Mg phase, and a small amount of MgO phase, and the microhardness ranged from 60 to 68HV. This hardness was greater than that of as-cast pure magnesium, and was attributed mainly to grain refinement and solid-solution strengthening. As the laser-energy was increased, the porosity decreased and the compression behaviour improved.

Friction stir additive manufacturing was applied[171] to Mg-4Y-3Nd and AA5083. In the case of the magnesium-based alloy, an average hardness of 120HV was found for the product; considerably higher than that (97HV) of the base material. In the case of the aluminium-based alloy, the average product hardness was 104HV as compared with the base hardness of 88HV. A multi-layered stack of magnesium-based WE43 alloy was built[172] using friction-stir additive methods, showing that the formation of defects was sensitive to the heat input and that dynamic recrystallization led to finer (2 to 3μm) grain sizes. The fine grain size, together with suitable precipitate behaviour, led to good mechanical properties. A maximum hardness of 115HV was found for the as-prepared material, and this increased to 135HV during aging. These figures were suggested to correspond to a strength of 400MPa and a ductility of 17%.

A study of laser additively manufactured porous ZK61 (Mg-5Zn-0.33wt%Zr) alloy showed[173] that the resultant surface morphology and porosity depended upon the laser energy. With increasing zinc content, the surface quality worsened, the grain structured was refined and the precipitate phases underwent successive transitions from MgZn, via MgZn plus $Mg_7Zn_3$, to just $Mg_7Zn_3$. The microhardness was markedly improved, and ranged from 57.67 to 109.36HV. This was attributed to Hall-Petch strengthening, solid-solution strengthening and precipitation-hardening. Samples having the composition, Mg-15Zn-0.3wt%Zr, possessed an ultimate compressive strength of 73.07MPa and an elastic modulus of 1.785GPa.

Wire arc methods were used[174] to produce parts from magnesium-based AZ31B alloy wire. The tensile strength of the finished object almost equalled that of conventional rolled stock. The average tensile strength of material prepared using the wire arc method was 239MPa; well within the specification for rolled stock: 221 to 275MPa. The observed 0.2% yield strength of the wire arc material was 95MPa; lower than that (200MPa) of rolled stock, while the wire arc material had an elongation of 21%; higher than that (12%) of rolled material. An advantage of the present process is that the strains introduced by the rolling process are absent.

Most recently, the use of fused filament fabrication has been proposed for the forming of complicated magnesium-based parts because this sintering technology does not require expensive moulds and can create low-cost prototypes. Magnesium-alloy powder-polymer blends have been formulated[175] which permit the manufacture of flexible filaments and robust green parts that can then be consolidated by sintering. Typical specimens have ultimate tensile strengths of up to 177MPa, a yield strength of 123MPa and an elongation-to-fracture of 2.8%. As these properties are as good as those of as-cast material, fused filament fabrication is deemed to be a viable method for the production of magnesium implants. It is known that varying the process parameters affects the ultimate tensile strength of fused filament fabricated parts, and that the scale of the component and the number of shells are the most important parameters which affect the strength. Further influences[176] are the specimen width, the specimen thickness, the number of shells and the in-fill density. There exists an inverse relationship between the part's scale, change in cross-sectional area and ultimate tensile strength. Further research[177] has examined the tensile strength of partially-filled fused filament fabricated parts with regard to the dimensions of solid floor/roof layers, shells and in-fills. Partially-filled fused filament fabricated parts consist of hollow sections. Because of the voids, the conventional method of determining the ultimate tensile strength from the gross cross-sectional area cannot be used and so a mathematical model which gave a more accurate representation of the cross-sectional area of a partially-filled component was derived. The model could also predict the dimensions and lateral distortions of the features of an additively manufactured part by using experimental data.

**Nickel**

Before considering the properties which result from the additive manufacture of the well-known nickel-based alloys, it is instructive to summarise the results obtained for a range of other nickel-based compositions.

Selective laser melting has been used[178] to process pure nickel powder. The scan-speed had a marked effect upon the porosity, and a relative density of 98.9% was obtained together with a microhardness of 140 to 160HV and a tensile strength of 452MPa. Samples of 1718 (Ni-19Cr-18Fe-3wt%Mo) were prepared[179] using powder-blown additive manufacturing, showing that the microstructure and mechanical properties were greatly affected by the process parameters. Fine equiaxed grains were found, and there was no change in microstructure along the build direction, thus imparting a uniform hardness of about 260HV. The principal material chosen for the laser additive manufacturing of aerospace-related turbine blades is K417G (Ni-10Co-9Cr-5.1Al-4.5wt%Ti) superalloy powder. When powder having a mean D50 particle size of 74µm, a flowability of 16.6s/50g, an apparent density of 4.78g/cm$^3$ and an oxygen content of 0.015% was prepared via vacuum induction-melting gas-atomisation, the product microstructure consisted of $\gamma$ and $\gamma'$, plus carbides. The average microhardness was

410HV, the tensile strength was 1080MPa, the yield strength was 828MPa and the elongation was 13.52%[180].

The nickel-based superalloy, K465 (11Co-10.5W-10Cr-5.5Al-2.8Ti-2.2wt%Mo), has been deposited onto AISI316L stainless steel by means of laser solid forming[181]. The as-deposited K465 consisted mainly of columnar dendrites which grew epitaxially from the substrate; the dendrites being fine with an average primary-dendrite arm-spacing of 18µm. Only a few fine equiaxed grains appeared at the top of the deposit. The $\gamma'$ particles in the deposit were of an essentially uniform spherical shape and were much finer than those which are found in as-cast K465. The average hardness of the as-deposited samples was 468HV; higher than that of as-cast K465. The tensile strength and the yield strength of as-deposited samples were 1205.4 and 917.4MPa, respectively, with an elongation of up to 8.5%.

Composites consisting of Ni-18Al-11Cr-9C and Ni-14Al-8Cr-29at%C (with the carbon present as graphite) were prepared[182] by means of laser engineered net shaping. Due to the variation in graphite content, differing microstructures of nickel aluminide and chromium carbide formed during solidification. This then affected the dry sliding-wear behavior. It was found that NiO and $Cr_2O_3$ were the main phases at the room-temperature surfaces while, at 500C, $NiCr_2O_4$ spinel structure was the main tribochemical phase: it formed a glazed oxide layer at the sliding surface, which lessened friction and wear. The increase from 9 to 29at%C did not decrease the dry-sliding friction coefficient, but there was a reduction in the 500C sliding wear rate to 7.1 x $10^{-6}mm^3$/Nm and an increase in the microhardness and macrohardness to $539HV_{0.3}$ and 48HRC, respectively.

It is difficult to produce bulk quantities directly from metallic glass, thus limiting its practical application, but easily procured $Ni_{82.2}Cr_7B_3Si_{4.8}Fe_3$ metallic-glass thin strips, 1.7mm x 0.04mm, could be used[183] to manufacture bulk metallic glass additively by means of ultrasonic bonding. Fully amorphous bulk nickel-based metallic glass could be produced by choosing the appropriate ultrasonic vibration energy, while leaving its thermal properties essentially unchanged. Most importantly, the hardness and elastic modulus were improved: the hardness to 8.55 from 8.21GPa and the elastic modulus to 134.10 from 81.74GPa. The overall results indicated that bulk metallic glass, when manufactured by ultrasonic bonding, could support a greater pressure than usual.

The NiTi alloy is well-known for its shape-memory behaviour, but also possesses a high ductility and excellent corrosion and wear resistance. In addition, it exhibits good biological compatibility. Laser metal deposition can be used to exploit these unique properties while creating complex-shaped components. Selective laser melting has been used[184] to produce high-quality super-elastic parts from NiTi shape memory alloy. This showed that the process parameters had to be optimized to an energy-input of 234J/mm$^3$ and a speed of 0.2m/s so that a dense product could be produced while minimizing impurity pick-up. The properties of additively manufactured NiTi, including the shape memory and pseudo-elasticity behaviours, compared well with those of conventionally processed NiTi. Even under optimum conditions however, there could be a shift in the phase transformation to higher temperatures following processing. More recent studies, performed using micro direct metal deposition of NiTi powder alloy, indicated that increasing the scanning speed improved the surface quality and decreased the micro-porosity content[185].

*Hastelloy X*

Laser powder-bed fusion additive manufacture of Hastelloy X parts showed[186] that a unidirectional-stripe scanning pattern led to the largest grain size, of more than 850µm, but also to the lowest strength. These samples possessed the strongest crystallographic texture, resulting in a planar anisotropic mechanical response, with a 22MPa difference in the ultimate tensile strength. Stripe rotation, by 67°, of the scan led to a randomly oriented and finer (110µm) grain structure with a higher (800MPa) ultimate tensile strength due to grain refinement. The aspect ratio of the columnar grain structure affected the mechanical response. The ultimate tensile strengths of horizontally printed parts were some 26% higher than those of the vertical parts for the 67° stripe-rotation specimens. Changing the solidification pattern by a 90° rotation reduced that difference to about 18%.

*Haynes-282*

Direct laser metal deposition has been used[187] to produce specimens of the Ni-Cr-Co alloy, Haynes-282, which were then solution heat-treated (1120C, 2h) before two-step aging (1010C, 2h plus 788C, 2, 4, 6, 8, 16 or 24h). Some samples were subjected only to the two-step aging or to one-step aging (788C). The average gamma-prime precipitate size in the microstructure of as-deposited samples was 14.5nm in the interdendritic region, plus 0.77% of carbides. Following aging, 14.5nm gamma-prime particles were found in the dendrite region plus 22.5nm particles in the interdendritic region, while the carbide content increased to 2.15%. One-step aging (788C, 24h) imparted a very good combination of room-temperature properties, with a yield stress of 950MPa, an ultimate tensile strength of 1240MPa and an elongation of 18%. In further research, a 100mm x 15mm x 25mm block of Haynes-282 superalloy was fabricated[188] by laser metal deposition. The use of post-deposition heat-treatments produced a wide range of strengths and ductilities, which could be adjusted so as to obtain the optimum set of properties. There existed columnar dendrites having, with a [001] growth axis, which were oriented opposite to the heat-flow direction during deposition. The as-deposited material contained $\gamma$-phases and molybdenum-rich $M_6C$, with the strengthening $\gamma'$ phase having an average size of 8.25nm in the interdendritic region. There was marked segregation of titanium and molybdenum in the interdendritic regions. The resultant hardness of 294HV, yield strength of 633MPa and 31.5% elongation of the as-deposited material were better than the corresponding as-cast properties. An as-deposited specimen which was heat-treated (788C, 16h) contained an homogeneous distribution of $\gamma'$, with an average size of 31nm plus globular and Chinese-script carbides in the interdendritic region. This specimen exhibited a hardness of 410HV, a yield strength of 894MPa, an ultimate tensile strength of 1200MPa and an elongation of 18%.

*Inconel-625*

During the selective laser melting of Inconel-625 alloy[189], the melt-pool contained elongated columnar crystals with, due to the rapid cooling-rate, a primary dendrite-arm spacing of about 0.5μm. The growth axis was usually <001>, due to epitaxial growth and heat conduction. The associated hardness was 343HV. There was a fully-formed austenite structure which exhibited large lattice distortions and no carbides or precipitated phases were present. Following heat treatment, the grains had grown into two microstructures having distinct differences: rectangular grains limited to the melt pool, and equiaxed grains lying along the molten boundaries. Large numbers of zig-zag grain boundaries were observed. Also following heat treatment, triangular MC carbides precipitated. The microstructural evolution of laser powder-bed additively manufactured material during subsequent stress-relief annealing (870C, 1h) led[190] to the formation of an appreciable fraction of the orthorhombic $Ni_3Nb$ $\delta$-phase which impairs the fracture toughness and ductility of conventional wrought material. The $\delta$-phase platelets precipitated within the interdendritic regions of the as-prepared solidification microstructure; the regions which were enriched in solute elements such as niobium and molybdenum due to micro-segregation during solidification. Precipitation of the $\delta$-phase at 800C took up to 4h, thus suggesting that alternative stress-relief processing could impede $\delta$-phase formation. Homogenization heat treatment was thus recommended for additively manufactured material.

The powdered metals used as feedstock for selective laser melting were studied in some detail with regard to their particle-size distribution, flowability, mechanical properties and microstructure[191]. The presence of any appreciable proportion of particles which were smaller than 10μm led to marked agglomeration and made difficult the selective laser melting process. Laser metal deposition has been used[192] to create Inconel-625 composites reinforced with TiC nanoparticles. A relatively low laser-energy input per unit length of 33kJ/m produced insufficient liquid of higher viscosity, hindered the melt from spreading out smoothly and led to the formation of a large number of micropores and a reduced densification of the part. When an optimum laser-energy input per unit length of 100kJ/m was used, the densification approached 98.8% due to the action of Marangoni flow within the melt pool. The carbide reinforcement underwent successive changes, from agglomeration to a uniform distribution, as the laser energy per unit length was increased. The columnar dendrites of the matrix were coarsened by a laser-energy input of 160kJ/m. Almost fully-dense parts exhibited a microhardness of $330HV_{0.2}$, a coefficient-of-friction of 0.41 and a dry-sliding wear-rate of 5.4 x $10^{-4}$ $mm^3$/Nm. The formation of fine columnar dendrites of Ni-Cr γ-phase, combined with homogeneously distributed ultra-fine reinforcing particles further led[193] to a coefficient-of-friction of 0.30 and a wear-rate of 1.3 x $10^{-4}mm^3$/Nm. Optimally-prepared composite parts had a tensile strength of 1077.3MPa, a yield stress of 659.3MPa and an elongation of 20.7%; with a ductile fracture mode.

*Table 22. Tensile properties of laser-aided additively manufactured Inconel-625 composites*

| Reinforcement | UTS (MPa) | 0.2%Yield Strength (MPa) | Elongation (%) |
|---|---|---|---|
| None | 840 | 531 | 16 |
| Graphite | 969.89 | 687.50 | 13.36 |
| Carbon nanotubes | 1005.55 | 695.11 | 21.44 |

These superior properties were attributed to grain refinement of the matrix during laser processing and to the blocking effect of ultrafine reinforcing particles on the mobility of dislocations. The incorporation[194] of carbide particles also changed the predominant texture of the Ni-Cr matrix, from (200) to (100). The bottom and sides of a deposited track consisted mainly of columnar dendrites, while cellular dendrites predominated in the central zone of the track. When carbide particles were added, more columnar dendrites appeared in the solidified melt pool and the incorporation of nanoparticles caused the formation of highly refined columnar dendrites having quite well-developed secondary arms.

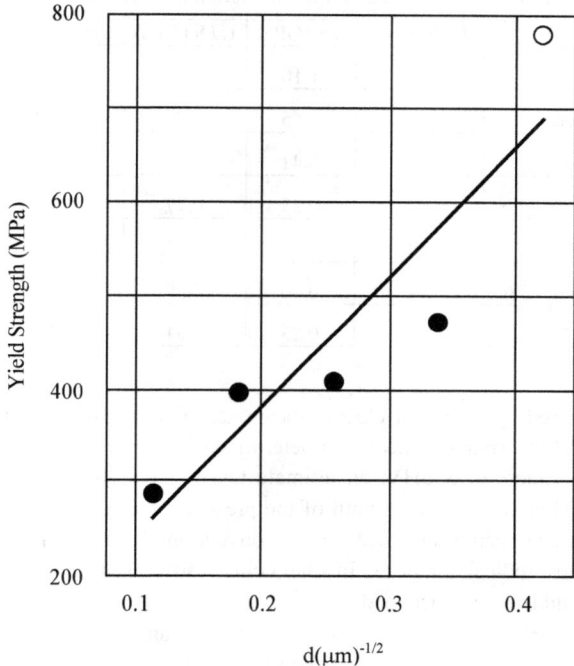

*Figure 20. Hall-Petch curve for Inconel 625, based upon conventionally prepared samples (black), with outlier (white) for additively manufactured material*

When microparticles were added, the columnar dendrites coarsened and degenerated, with secondary growth being completely suppressed. When nanoparticles reinforced the Inconel 625, a greatly improved microhardness and wear were obtained without loss of ductility. A comparison has been made[195] of the use of carbon nanotubes and graphite to produce composites from Inconel 625 via laser-aided additive manufacturing (table 22). The nanotubes were found at columnar dendrite boundaries, together with precipitated Laves and $\gamma'$-phases. The addition of 0.25wt% of carbon nanotubes could increase the strength and ductility by 20%, due to grain refinement and grain-boundary pinning. The use of graphite increased the strength but decreased the ductility, due to the increase in carbon content.

*Table 23. Mechanical properties of electron-beam melted and wrought Inconel-625*

| Material | Rockwell-C | YS (GPa) | UTS (GPa) | Elongation (%) |
|---|---|---|---|---|
| wrought, cold-worked | 40 | 1.10 | | 18 |
| wrought, annealed | 20 | 0.45 | 0.89 | 44 |
| EBM, as-prepared | 14 | 0.41 | 0.75 | 44 |
| EBM & HIP | 8 | 0.33 | 0.77 | 69 |
| Wrought, 538C | 18 | 0.28 | 0.83 | 50 |
| EBM, as-prepared, 538C | 14 | 0.30 | 0.59 | 53 |
| EBM & HIP, 538C | 6 | 0.23 | 0.61 | 70 |

Inconel 625 was prepared by means of electron-beam selective melting, and its fretting wear against the 42CrMo4 stainless steel was determined[196] during flat-on-flat contact. The material had a hardness of 335HV, an ultimate tensile strength of 952MPa and a yield strength of 793MPa. The yield strength of the present material did not fall on the Hall-Petch line which had been established for the conventionally prepared alloy (figure 20). The tribological test indicated that the Inconel could restrict wear of the surface to a depth of just 2.4µm, under a contact load of $10^6$N, after 2 x $10^4$ cycles. The excellent wear-resistance was attributed to the improved strength and to the formation of continuous layers containing a mixture of $Fe_2O_3$, $Fe_3O_4$, $Cr_2O_3$ and $Mn_2O_3$. The relationship between the beam current, speed and focus and the elastic modulus and yield stress of additively manufactured mesh cubes was such that the moduli and yield stress could be varied by a factor of about ten by changing the electron-beam parameters. Simple models were found to explain the relationships[197]. Anisotropy of the mesh was associated with the layered structure and could contribute, together with microstructural anisotropy, to the anisotropic mechanical properties of the mesh. The properties of as-prepared and heat-treated Inconel-625 samples, produced via the electron-beam melting of pre-alloyed precursor powder, were compared[198] for untreated and hot-isostatically pressed (1120C) cylinders. There was an initial electron-beam melting-developed γ″ $Ni_3Nb$ precipitate platelet columnar structure within columnar [200]-textured γ Ni-Cr grains which were aligned along the cylinder axis, parallel to the electron-beam build-direction. During annealing (1120C), the precipitate columns dissolved and the columnar γ grains recrystallized to form generally equiaxed grains with coherent {111} annealing twins, containing $NbCr_2$ Laves-phase precipitates. The micro-indentation hardness decreased from about 2.7 to about 2.2GPa following hot-isostatic pressing, and the corresponding 0.2%-offset yield stress decreased from 0.41 to 0.33GPa, while the ultimate tensile strength increased from 0.75 to 0.77GPa. The corresponding elongation increased from 44 to 69% in the case of hot isostatically pressed components (table 23).

In order to compare the currently competing powder-bed based additive manufacturing techniques, Inconel-625 powder was processed[199] using electron-beam powder-bed laser fusion, laser powder-bed fusion and binder-jetting. The samples were made in X and Z build directions, and all of them were subjected to subsequent hot isostatic pressing. Microstructural examination revealed equiaxed grain formation in the case of binder jetting and laser powder-bed fusion material, while the other samples had columnar grains parallel to the build direction. The results (table 24) showed that all 3 methods could satisfy minimum specification requirements in the case of parts built in the X direction. In the case of the Z direction, however, only laser powder-bed fusion material could satisfy minimum standard requirements. In fact, that technique was the better one with regard to most properties, followed by electron-beam powder-bed laser fusion and binder jetting. Ductile fracture with dimples occurred in specimens produced by using any of the methods. Some non-sintered powder particles survived during binder jetting, and a so-called woody structure existed in laser powder-bed fusion material.

*Table 24. Properties of Inconel-625 powder processed using competing techniques*

| Technique | Direction | YS (GPa) | UTS (GPa) | E (GPa) | Elongation (%) |
|---|---|---|---|---|---|
| binder jetting | X | 0.32 | 0.707 | 0.542 | 58.74 |
| binder jetting | Z | 0.393 | 0.708 | 0.506 | 27.02 |
| electron-beam | X | 0.367 | 0.849 | 0.484 | 44.32 |
| electron-beam | Z | 0.369 | 0.723 | 0.459 | 26.92 |
| laser powder-bed | X | 0.396 | 0.906 | 0.561 | 62.34 |
| laser powder-bed | Z | 0.349 | 0.842 | 0.539 | 56.3 |

In powder-bed binder jet printing, powder is deposited layer-by-layer and selectively joined using binder at each layer. The powder does not melt during printing, so the density following printing is only about 50% and sintering is required in order to finish the as-printed component. The method has been used[200] to produce parts made from Inconel-625 and, in order to deduce the optimum sintering temperature, the as-received powder was subjected to differential scanning calorimetry analysis and printed green samples were cured and sintered at various temperatures under a high vacuum. Fully dense parts having densities of up to 99.6%, a hardness of up to $238HV_{0.1}$ and an ultimate tensile strength of up to 612MPa, could be obtained by sintering at 1280C. It was concluded that this alloy, when produced using powder-bed binder jet printing, could have a similar density and mechanical properties to those resulting from conventional casting. The surface roughness and near-surface porosity of binder-jet additively manufactured Inconel-625 parts was studied[201] using an electrospark deposition technique. Localized surface melting and material transfer from the electrospark deposition electrode produced a near-surface region of increased density (from 62.9 to 99.2%) and an increased hardness (from 109 to 962HV). An *in situ* method for metal-matrix formation has been proposed[202] which involves a binder jet additive manufacturing process with just Inconel-625 powder and a carbon-containing binder. Carbide formation then occurs during high-temperature sintering due to the combination of chromium, molybdenum and niobium from the Inconel with carbon from the binder. The core-shell morphology and composition of the composite could be controlled by adjusting the amount of carbon in the system during sintering.

Friction stir additive manufacturing was applied[203] to Inconel-625. Grain-refinement occurred during layer deposition, with fine equiaxed grain structures. Grains as fine as 0.27μm were observed in the interface regions, while the average grain size was about 1μm. The strain-rate dependence was determined by using quasi-static (0.001/s) and high (1500/s) rate tensile experiments involving a direct tension-Kolsky bar. The high-speed results revealed an approximately 200MPa increase in strength over the quasi-static results. The fracture surfaces at both strain-rates were aligned with the maximum shear plane, and exhibited localized micro-voids.

When this alloy was processed using cold metal transfer wire arc additive manufacturing[204], microstructural studies revealed variations in the microstructure of the various layers of the specimen, with the bottom layer comprising fine primary cellular grains. With increasing travel speed, the average microhardness of manufactured specimens improved slightly, from 248 to 253HV while the ultimate tensile strength increased from 647 to 687MPa and the yield stress increased from 376 to 400MPa. This mechanical performance, apart from the ultimate tensile strength, was better than that of cast Inconel-625.

*Inconel-718*

In early additive manufacturing research, Inconel-718 was manufactured[205] directly by laser deposition using the optimized parameters of a laser power of 800W, a laser beam diameter of 0.8mm, a scanning speed of 0.5m/min and a powder feed-rate of 3g/min. The microstructures of laser-deposited samples comprised directionally solidified columnar features which were metallurgically bonded to the substrate. The average hardness was about $440HV_{0.2}$. The tensile strength was 121 and $116kg/mm^2$ at room temperature and at 650C, respectively. These values were a little lower than those (142 and $127kg/mm^2$) for forged Inconel-718 plate. This was attributed to its directionally solidified columnar structure being perpendicular to the tensile-testing load. Selective laser melting processed Inconel-718 was examined[206] following heat treatment. An obvious weld-beads structure could be observed on the unheated XY surface of the alloy and the microstructure exhibited serious inhomogeneities, with a distribution of columnar crystals and fine dendrites. With increasing solid-solution temperature, the microstructure at the alloy surface exhibited clear refinement. The surface hardness also gradually decreased. The usual columnar crystals, dendrites, intermetallic compounds and precipitation-hardening phases (FeNi, $Ni_8Nb$) were also reduced in number or not present. Below the solid-solution temperature of 950C, fine homogeneous $\delta$ and $\gamma'$ phases in or near to the grain boundaries had a marked effect upon the XY surface hardness, which ranged from 476 to 500HV

In as-processed Inconel-718 alloy, prepared by selective laser melting, ultrafine columnar grains with highly dispersed $\gamma''$ precipitates at grain boundaries, and an even distribution of $\gamma'$ phase within grains, were found[207]. Longitudinal as-prepared samples had an ultimate tensile strength of 1101MPa and an elongation of 24.5%, as compared with transverse samples which had an ultimate tensile strength of 1167MPa and an elongation of 21.5%. The good mechanical properties for both orientations was attributed to the fine microstructure which in turn resulted from the high cooling-rate that was caused by the laser processing. Any anisotropy in strength and ductility was attributed to the {100} fiber texture and to the columnar grain morphology. The {100} fiber texture of the columnar grains led to there being a high strength in the transverse direction, but the columnar grains also offered a path along which damage could accumulate and lead to fracture.

The effects of heat-treatment upon as-prepared laser powder-bed additively manufactured Inconel-718 were investigated[208]. In as-prepared material, columnar grains with interdendritic microsegregation formed along the build direction, with equiaxed grains forming on sections perpendicular to the build direction. Following prolonged soaking during simulated hot isostatic pressing, the microstructure changed from heterogeneous columnar grains to homogeneous recrystallized grains with MC-type precipitates. This led to a change in microhardness from 281 to $171HV_{2.0}$, while the Young's modulus changed from 209 to 229GPa. The effect of aging in increasing both hardness and modulus was attributed to the formation of $\gamma''$ and $\gamma'$ precipitates in the nickel matrix, and to the small effective grain size.

A composite consisting of an Inconel-718 matrix and 0.5wt% nano-TiC reinforcement was prepared[209] by using selective laser melting. It and plain Inconel-718 were then solutionized at 940, 980 or 1020C for 1h, annealed at 1100C for 1h and subjected to two-step aging (720C, 8h, furnace-cool and 620C, 8h, air-cool). Compared with the plain Inconel-718 prepared using selective laser melting, the Inconel composite had a better ultimate tensile strength in both the as-prepared and solutionized or annealed conditions. The dendritic structure of the Inconel was markedly refined by the carbide particles in as-prepared samples, and grain-coarsening was largely inhibited by the carbide particles in solutionized or annealed samples. In both reinforced and plain Inconel, dissolution of Laves phases and the precipitation of δ-phase were observed. Annealing at 1100C was not favorable to the formation of δ-phase. Aging markedly increased the ultimate tensile strength of both types of material, but the strengthening effect of added nano-particles became less significant in the aged condition due to the precipitation of $\gamma'$ and $\gamma''$. In related work[210], the Inconel was reinforced with graphene nanoplatelets and the composite was again prepared using selective laser melting. The pure Inconel, and graphene-reinforced Inconel comprising 0.25 or 1wt% of platelets, were compared. The selective laser melting process ensured an even dispersion of the graphene in Inconel powders. The resultant composites had a dense microstructure and a greatly improved tensile strength. The ultimate tensile strengths were 997.8, 1296.3 and 1511.6MPa, and the Young's moduli were 475, 536 and 675GPa, for 0, 0.25 and 1wt% of platelet reinforcement, respectively. The bonding between the platelets and the matrix appeared to be strong, and the platelets were retained well during laser-melting. It was deduced that load-transfer, coefficient of thermal expansion mismatch and dislocation-blocking were the main reinforcing mechanisms. In later work[211], the composites were solution heat-treated (980 or 1220C, 1h) and two-step aged. In the as-prepared condition, the ultimate tensile strengths were 997 and 1447MPa, respectively, for 0 and 4.4vol% of reinforcement. The strengthening effect was most marked in the as-prepared condition. Precipitation-hardening by $\gamma$ and $\gamma'$ phases was suppressed in the aged material due to the formation of MC carbide and a depletion in niobium.

The distribution of pore sizes in laser-deposited Inconel-718 was investigated[212] by using laser-powers of up to 5.5kW and deposition rates of up to 7000mm³/min. Using quantitative image analysis and logarithmic Gaussian distributions, the pore-size distribution in the bulk material could be described by using just 4 statistical parameters. The ultimate tensile strength of a clad sample could be estimated with an accuracy of 250 to 300MPa.

The effects of build direction and heat treatment, upon the tensile properties of samples which were prepared by direct metal laser sintering, were determined[213] at room temperature and 650C. Particle-like δ-phase was localized in the interdendritic regions due to a niobium segregation which occurs during direct metal laser sintering solidification. Vertical and horizontal specimens were taken parallel to, and perpendicular to, the build direction, respectively. The tensile strengths of heat-treated samples produced by direct metal laser sintering were comparable to those of cast and wrought samples. When the horizontal and vertical specimens were subjected to solution treatment and aging at 650C however, the horizontal specimens had one quarter of the ductility of the vertical specimens. This was due to the interdendritic δ-phase precipitates, which were arranged perpendicular to the stress axis in the former case. A row of interdendritic δ-phase precipitates with incoherent interfaces could affect the tensile ductility at high temperatures.

The tensile, creep, stress-rupture and notch-rupture properties at 650C of samples which had been prepared by laser powder-bed fusion were generally similar to those of the traditional wrought alloy[214]. An exception was that of the notch-rupture data, where all of the samples failed at the notch, with no elongation. These results were not sensitive to solution treatments performed above or below the δ-phase solvus temperature. This again was not characteristic of wrought material.

The metal additive manufacturing process known as laser engineered net shaping was used[215] to repair internal defects in specimens of the alloy. The technique involved the milling of slots around the defect zone and their re-refilling with new material. The effectiveness of slots having rectangular or trapezoidal shapes in repairing internal cracks was investigated, together with the effect of deposition-orientation upon the mechanical properties of a repaired component. A trapezoidal shaped cross-section led to good fusion between the deposited material and the component, with no void inclusion. The build-direction, especially in a diagonal sense, affected the coefficient of friction and the wear behaviour of non heat-treated samples. A model has been developed[216] in order to predict the microstructural evolution which occurs during this net shaping process. A detailed transient thermal finite-element method was combined with a density-type microstructural model which permitted calculation of the phase-fractions: Widmanstatten, basket-weave, alpha lath. Reasonable agreement with experiment was found.

It is to be noted that Laves-phase segregation is a universal occurrence in additively manufactured or as-cast Inconel-718 alloy. The presence of this phase usually impairs the mechanical properties of the matrix. When nickel-based alloys having various niobium contents were prepared[217] using laser metal deposition and a dual-feed system, the Laves-phase distribution at the macro level was relatively uniform. The grain size of the as-prepared nickel-based alloys decreased with niobium content, and the Laves phase always had a higher nanohardness but a lower surface potential than that of the $\gamma$-phase matrix. The nanohardness of the Laves phase increased linearly, while the potential decreased with the niobium content. The strength improvement of the as-prepared nickel-based alloys, with increasing niobium content, was attributed to grain refinement and to increases in the hardness and proportion of the Laves phase. An increase in the mainly $\gamma''$ precipitate content as a function of the niobium content led to a marked improvement in the hardness of heat-treated nickel-based samples, but the grain refinement strengthening was comparatively minor after aging.

Laser engineering net shaping had previously been used[218] to make bimetallic structures from Inconel-718 and GRCop-84 copper alloy, either via the direct deposition of the later alloy on Inconel-718 or via compositional gradation of the alloys. The latter option produced a gradual transition of Inconel-718 and GRCop-84 elements at the interface, as revealed by the cross-sectional hardness profile across the bimetallic interface. There were columnar grain structures at the interface, with an accumulation of $Cr_2Nb$ precipitates along grain boundaries and the substrate/deposit interface. The mean thermal diffusivity of the bimetallic structure was $11.33mm^2/s$ between 50 and 300C, as compared with the diffusivity ($3.20mm^2/s$) of pure Inconel-718. The conductivity of the bimetallic structures was also increased by some 300% as compared with that of the Inconel.

Selective laser melting has been used[219] to prepare titanium carbide reinforced Inconel-718 nanocomposites. An insufficient laser energy-density input limited the densification of shaped parts, due to the formation of large pore chains or interlayer micropores. The densification increased to a nearly complete level as the applied energy-density was optimized. The carbide reinforcement generally changed, from being highly agglomerated into polygonal shapes to being uniformly distributed with smooth fine structures, upon increasing the applied energy-density. The columnar-dendritic matrix meanwhile exhibited marked epitaxial growth. The optimally prepared fully-dense material exhibited a high microhardness, with a mean value of $419HV_{0.2}$, a friction coefficient of 0.29 and a wear-rate of $2.69 \times 10^{-4}mm^3/Nm$ during dry-sliding. In a further study, TiC nano-particles were used[220] to reinforce Inconel-718 and the mixed raw powder was then processed using the selective laser melting technique. Samples with TiC contents of 0, 0.4, 0.8 or 1.6wt% were subjected to standard solid-solution treatment (980C, 1h). An increase in the reinforcement content led to a higher tensile strength. With 1.6wt% of reinforcement, the ultimate tensile strength increased by 15%. The composites had their maximum yield stress and ultimate tensile strengths when in the as-prepared condition while, in the solution-treated condition, the tensile strengths were generally lower due to microstructural coarsening. In both as-prepared and solution-treated conditions, the load-effect strengthening was very small when compared with other contributions. Thermal mismatch strengthening was the most important factor at any volume fraction, for as-prepared samples. This was due mainly to the high selective laser melting temperature. In solution-treated samples, the mismatch strengthening diminished because the solution treatment largely relieved the thermal strain and eliminated most of the strain-induced dislocations. Hall-Petch strengthening became predominant because the large numbers of nanoparticles effectively inhibited grain coarsening during solution treatment. When nano-TiC reinforced Inconel-718 composites were most recently prepared by means of selective laser melting[221], a fully dense composite part could be fabricated by using a laser-energy linear density of 300J/m. Upon increasing the energy-density from 225 to 300J/m, the reinforcement underwent severe agglomeration along the grain boundaries and within the matrix grains. The morphology of the nano-particles changed from irregular-polygonal to near-spherical. The presence of nano-TiC could accelerate the refinement of the columnar-dendrite spacing of the $\gamma$-phase matrix. A nanohardness of 4.48GPa, a coefficient-of-friction of 0.36 and a wear-rate of 3.83 x 10$^{-4}$mm$^3$/Nm were obtained by using an energy-density of 300J/m.

Electron-beam melting was used to produce Inconel-718 components which were then subjected to various treatments[222]. The average yield strength increased from 980 to 1180MPa, the ultimate strength increased from 1160 to 1350MPa and the microhardness increased from 410 to 470HV, due to heat-treatment. The corresponding elongation-to-fracture however decreased from 8.2% in the as-processed condition to 6.5% in the heat-treated condition. The electron-beam melting process was used to produce[223] Inconel-718 specimens from a single batch of virgin plasma-atomized powder. Typical columnar microstructures were observed. Some of the specimens were subjected to solution-treatment and aging, while another set was subjected to hot isostatic pressing; followed by the above treatment. Most of the treatments led to a hardness increase of the order of 50HV (table 25). Fractography of tensile-tested specimens showed that closure of shrinkage porosity and the partial healing of non-fused regions led to the improved properties.

*Table 25. Tensile properties of electron-beam processed and heat-treated Inconel-718*

| Condition | E (GPa) | YS (MPa) | UTS (MPa) | Elongation (%) |
|---|---|---|---|---|
| as-processed | 138 | 920 | 1075 | 10 |
| solution/aged | 137 | 1096 | 1172 | 6 |
| HIP & solution/aged | 142 | 1100 | 1190 | 14 |

Electron-beam melted Inconel-718 has been solution-treated and aged by heating the top surface of the product so as to act as a planar heat source during cooling[224]. The hardness of material subjected to such an *in situ* heat treatment was 478HV; similar to that of peak-aged Inconel 718. Large grains and cracks formed and were a likely cause of failure, during tensile tests, of the *in situ* heat-treated material. Ignoring the poor tensile behaviour, this method could increase precipitate size and hardness while reducing the number of processing steps.

By using a point heat source for the electron beam melting process of additive manufacturing it is possible[225] to form columnar or equiaxed grain structures during solidification by adjusting the parameters which are associated with the point source. The material may then exhibit either anisotropic properties, in the case of columnar-grained material made using a standard raster scan, or isotropic properties in the case of equiaxed material made using the point heat-source.

High-strength materials having a hardness greater than 400HV are affected by the presence of small defects, and a disadvantage of additive manufacturing is unfortunately that it tends to introduce such defects. In Inconel-718 fatigue specimens with a hardness of 470HV, the defects were mainly gas porosity and non-fused areas[226]. Because the defect orientation in additively manufactured material is random, those defects which intersect the specimen surface have a greater effect. The area parameter model was found to be applicable[227]. Although the usual statistical analysis was valid for the quality control of additively manufactured components, the surface effect upon the effective magnitude of the defect size had to be carefully considered. Because the orientations of defects in additively manufactured materials are random, those defects in contact with the specimen surface have a greater effect upon the fatigue strength than does an internal defect and have a larger effective size than the real size, in the context of fracture mechanics.

The effect of oxides, wire source and heat treatment upon the properties of wire arc additively manufactured Inconel-718 have been studied[228], showing that oxides which formed during deposition had no effect upon the mechanical properties because a 0.5µm-thick passivating layer of $Cr_2O_3$ and $Al_2O_3$ formed during deposition and prevented further oxides from forming within the bulk. Using wires from various sources could lead to 50MPa difference in ultimate tensile strength. This was attributed to slight compositional variations and TiN inclusions. Standard heat-treatments could increase the strength from 824 to 1110MPa in the horizontal direction, but the average strength was 105MPa lower than that of the wrought alloy. As deposited material contained large columnar grains and numerous Laves phase, as compared with the fine grains of laser powder-bed fused or wrought Inconel-718. This initial microstructure led to the formation of less favourable and less numerous precipitates during heat treatment, leading in turn to the strength difference.

Compositionally optimized superalloy powder for additive manufacturing was prepared[229] by means of vacuum induction melting inert-gas atomization, and sieving. Following sieving, the mean particle size of the powder was less than 35μm and the particle-size distribution ranged from 10 to 55μm. The powder exhibited the high flowability which is needed for selective laser melting additive manufacturing. Tensile-testing of the resultant product at room temperature and 650 showed that the high-temperature properties were better than those of commercial Inconel-718.

Samples of the superalloy were prepared[230] by using the powder-bed fusion process, with 5 different sets of process parameters producing a range of porosities. The energy-absorption was very sensitive to the density, and the stress-strain response during compression-testing was much like that of open-cell foams in the presence of high porosity. The similarity to open-cell foam was confirmed by scanning electron microscopy, and micro computed-tomography scans indicated that the induced porosity was continuous throughout the material.

An attempt was made[231] to transfer the knowledge gleaned from traditional casting and welding, concerning the correlation between solidification microstructure and mechanical properties, to additive manufacturing; especially with regard to the repeated melting and solidification which occurs during electron beam powder melting. A melt-scan strategy was developed for the electron-beam melting of Inconel-718, and the 3-dimensional heat-transfer conditions were modelled numerically. Spatial and temporal variations in the temperature-gradient and growth velocity at the liquid/solid interface of the melt pool were predicted as a function of the electron-beam parameters. Analysis of the parameters furnished the optimum processing conditions that would result in a columnar-to-equiaxed transition during solidification, as confirmed by experiment.

*Table 26. Properties of as-deposited and heat-treated laser solid formed Inconel-738LC*

| Sample | Yield stress (MPa) | UTS (MPa) | Elongation (%) |
|---|---|---|---|
| as-deposited | 861 | 1157 | 8.8 |
| heat-treated (1070C) | 858 | 1224 | 9.9 |
| heat-treated (1120C, SHT*) | 900 | 1177 | |
| heat-treated (1160C) | 799 | 1158 | 3.6 |

*Standard heat treatment: 1120C, 2h, air-cool plus 845C, 24h, air-cool

## Inconel-738

Induction-assisted laser solid formed Inconel-738LC was subjected[232] to various solution treatments, at 1070, 1120 and 1160C, with the results showing that both as-deposited and heat-treated deposits predominantly contained columnar grains having a width of 100 to 270μm. Equiaxed grains arose mainly from the columnar-to-equiaxed transition during laser solid forming, while a small fraction of equiaxed grains arose from recrystallization during heat treatment, due to the relatively low temperature-gradient obtaining during the induction-assisted forming. Fine-grain strengthening still occurred in the columnar-grain dominated deposits, but γ'-phase strengthening played a large part in strengthening. Deformation of coarse (>200nm) near-cubic γ' helped to increase the ultimate tensile strength and elongation, while a brittle continuous phase which formed at the grain boundaries impaired the plasticity. The best compromise between tensile strength and plasticity was obtained by solution-treatment at 1070C (table 26).

*Table 27. Particle and tensile properties of wire arc processed Monel K500*

| Condition | DR | ND | Size (nm) | HV | YS (MPa) | UTS(MPa) | El. (%) |
|-----------|-----|-----|-----------|-----|----------|----------|---------|
| AP | 300 | 3.6 | 437 | 144 | 170 | 430 | 47 |
| AP | 400 | 3.1 | 448 | 141 | 165 | 410 | 51 |
| AP | 500 | 2.6 | 331 | 168 | 160 | 408 | 50 |
| A, 610C | 300 | 865 | 90 | 256 | 250 | 522 | 39 |
| A, 610C | 400 | 898 | 75 | 255 | 300 | 615 | 37 |
| A, 610C | 500 | 692 | 85 | 262 | 290 | 609 | 32 |
| A, 610&480C | 300 | 448 | 160 | 259 | 320 | 536 | 12 |
| A, 610&480C | 400 | 667 | 90 | 236 | 250 | 563 | 34 |
| A, 610&480C | 500 | 641 | 95 | 265 | 280 | 622 | 31 |

AP: as-processed, A: aged, DR: deposition-rate (mm/min), ND: number density ($10^{-3}/\mu m^2$)

Wire-arc additive manufacturing was used[233] to prepare ATI718Plus superalloy (Ni-19Cr-9Co-9Fe-5.5wt%Nb) specimens. The microstructures contained extensive eutectic constituents including Laves phases and MC-type carbides due to micro-segregation of elements such as niobium. Although the latter was reputed to promote the formation of δ-phase, only η-phase particles were observed. A predominant texture existed in <100> directions, with just a few misoriented grains found at the substrate/deposit boundary and on the top of the deposit. A 2mm soft zone was present, at the heat-affected zone, and was potentially harmful to the mechanical properties. This heat-affected zone also unsurprisingly contained the lowest number density of $\gamma'$ and $\gamma''$ precipitates. This was attributed to the diffusion of solute elements from the interdendritic regions.

*Monel K500*

Wire arc methods were used[234] to deposit a nickel-copper alloy, Monel K500 or FM60, in the form of a plate onto a metal substrate at a rate of 300, 400 or 500mm/min. The samples were then annealed (1100C, 0.25h, slow cool to 610C) and aged (610C, 8h) before being air-cooled to room temperature or slow-cooled to 480C, aged (480C, 8h) and air-cooled to room temperature. The precipitate size and number-density was correlated with the mechanical properties (tables 27 and 28). In Monel K500, the precipitates were mainly of TiC/TiCN type while, in FM60, they were of MnS and TiAlMgO type. The Monel K500 exhibited a greater hardness, strength, toughness and wear resistance. Aging at 610C improved the properties of both alloys, due to the precipitation of additional particles. Aging at 480C could however result in impaired properties if the particles coarsened.

Additive manufacturing methods have difficulty in producing walls which are less than 0.5mm thick. One approach is firstly to form a polymer template by additive manufacture, and then produce metal layers by using electroplating or chemical vapor deposition. Following removal of the polymer, the metal layers can be interdiffused to form an alloy by using homogenization heat-treatments. This process has been applied[235] to Monel alloys based upon Ni-Cu-Al-Ti, to nickel superalloys based upon Ni-Cr-Al and to rhenium alloys based upon Re-Co. The rhenium and cobalt were co-deposited by aqueous electroplating, resulting in an alloy with a melting point of 2000C and a Vickers hardness of $480HV_{0.2}$ after homogenization.

During additive manufacturing, the multiple thermal cycles which inevitably occur during the addition of subsequent layers lead to complicated transport phenomena and solidification behaviors in the melt pool, which then markedly affect the microstructure and mechanical properties. A three-dimensional numerical model was therefore developed[236] which took account of the thermal behavior, Marangoni effect, compositional transport, solidification and dendrite growth during the multi-layer additive manufacture of a nickel-based alloy on cast iron. Dimensional analysis simplified the force balance equation at the liquid/gas interface which governed the profile of the melt pool, and conservation equations of mass, momentum, enthalpy and concentration were solved in parallel. The solidification parameters at the liquid/solid interface in particular were deduced so as to be able to predict the solidification microstructure. The results showed that the cooling-rate progressively decreases with increasing numbers of deposited layers, thus leading to the occurrence of coarser solidified grains in the upper of part. Although powder and substrate could be mixed in an homogeneous melt pool, non-uniform concentration distributions persisted at the bottom of the deposit.

*Table 28. Particle and tensile properties of wire arc processed FM60*

| Condition | DR | ND | Size (nm) | HV | YS (MPa) | UTS(MPa) | El. (%) |
|-----------|-----|------|-----------|-----|----------|----------|---------|
| AP | 300 | 11.4 | 311 | 131 | 146 | 356 | 48 |
| AP | 400 | 7.2 | 392 | 132 | 149 | 361 | 47 |
| AP | 500 | 4.7 | 388 | 134 | 160 | 375 | 48 |
| A, 610C | 300 | 12.7 | 278 | 163 | 160 | 397 | 41 |
| A, 610C | 400 | 8.5 | 274 | 164 | 155 | 410 | 43 |
| A, 610C | 500 | 8.1 | 236 | 162 | 205 | 428 | 36 |
| A, 610&480C | 300 | 9.2 | 285 | 139 | 115 | 358 | 48 |
| A, 610&480C | 400 | 8.9 | 288 | 192 | 170 | 490 | 39 |
| A, 610&480C | 500 | 7.7 | 261 | 163 | 190 | 428 | 40 |

AP: as-processed, A: aged, DR: deposition-rate (mm/min), ND: number density $(10^{-3}/\mu m^2)$

## Niobium

For the purposes of powder-based additive manufacturing, spherical Nb-37Ti-13Cr-2Al-1Si pre-alloyed powder was prepared[237] by means of plasma rotating-electrode processing. The main phases present were niobium solid solution and $Cr_2Nb$. Fine dendritic structures were found in as-prepared pre-alloyed powders, and these transformed into large grains during heat treatment (1450C, 3h). With increasing powder size, the secondary dendrite-arm spacing increased and the microhardness decreased.

## Tantalum

Angular tantalum powder having low (87ppm) or high (829ppm) oxygen contents was spheroidized under argon by using radio-frequency plasma techniques, and the particle-size distributions were closely controlled[238]. Tensile test specimens were laser-printed under argon, with the results showing that low-oxygen spherical powder produced parts which exhibited higher elongations and ultimate tensile strengths and signs of ductile fracture. The highest (676MPa) ultimate tensile strength was found for a high-oxygen sample and was associated with a yield strength 587MPa, an elongation of 17% and a hardness of 296HV. The low-oxygen material exhibited an ultimate tensile strength of 634MPa, a yield strength of 601MPa, an elongation of 35% and a hardness of 149HV. When the oxygen content was below 100ppm, the elongations were similar to that of wrought tantalum. It was noted that the modulus of the additively manufactured material could be adjusted so as to match that of bone.

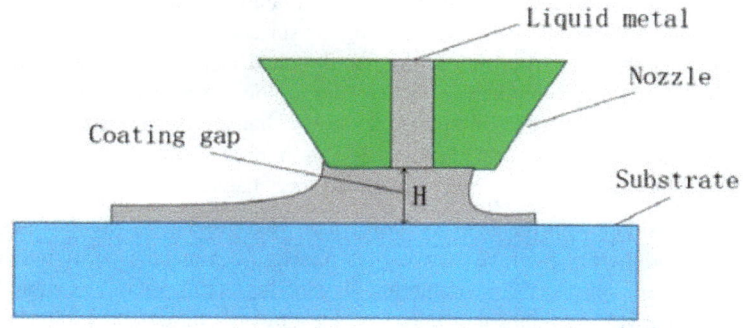

*Figure 21. The micro-coating technique of additive manufacturing. Reproduced from "Additive manufacturing of Sn63Pb37 component by micro-coating", Zhao, G., Wei, Z., Du, J., Liu, W., Wang, X., Yao, Y., Procedia Engineering, 157, 2016, 193-199 under Creative Commons Licence.*

## Tin

Parts made from Sn-37wt%Pb have been built in an induction furnace by using a nozzle to build them up on a copper-clad substrate (figure 21). Mechanical testing showed that the tensile strength of these printed parts was better by 20.4 and 11.9%, in the horizontal and vertical directions respectively, as compared with that of conventionally cast parts[239]. A crucible and a nozzle, instead of a welding torch and wire-feeder, can be used to supply material for shaped metal deposition in additive manufacturing. When applied[240] to Sn-37wt%Pb model alloy, the ultimate tensile strength perpendicular to the deposited layers was 40.89MPa and that parallel to the layers was 43.14MPa, with the respective yield strengths being 34.28 and 35.23MPa. These properties were superior to those of the conventionally cast alloy, for which the ultimate tensile strength and yield strength were 36.51 and 29.25MPa, respectively. This micro-coating method is very similar in fact to a suite of techniques which has been used for hundreds of years[241]. The difference is that the old methods sought only to produce a long thin even sheet of lead alloy. This would then be made into church-organ pipes, for example. In the present case, the deposit serves only as the base for a further deposit.

**Titanium**

In order to improve the wear-resistance of pure titanium which had been prepared by electron beam melting additive manufacturing, its surface was hardened[242] by using gas-carburizing. Following carburization, the hardness had increased and titanium carbides were found to have precipitated in the surface region. The friction coefficients were determined by using dry-friction sliding tests, and that of the carburized titanium was lower than that of the as-prepared material. This improvement in wear behaviour was attributed to the high surface hardness and the slippery nature of the titanium carbide.

Reactive-deposition additive manufacturing has been used[243] to create metal-matrix composites and thereby improve the wear-resistance of commercial-purity titanium. The composites were produced by promoting *in situ* high-temperature reaction between commercial-purity, zirconium and boron nitride powders during laser directed-energy deposition 3-dimensional printing. The commercially-pure titanium was printed with pre-mixed additions of 20wt%Zr and 10wt% BN by using laser engineered net shaping. There were TiN, TiB and $TiB_2$ reinforcing phases present in as-printed BN-containing structures. The combined addition of zirconium and boron nitride produced a Ti-Zr matrix which contained BN particles and an *in situ* phase-reinforced microstructure which had a 450%-improved hardness (from 318 to 1424HV) and a 9 times lower wear-rate as compared with that of similarly prepared titanium. The zirconium-addition alone led to a 12% greater hardness, a 23% higher compressive yield stress and an 11% lower wear-rate as compared with those of similarly-produced titanium.

The additive manufacturing of bulk Ti-TiC nanocomposite parts has been carried out[244] by using selective laser melting, and either ball-milled Ti-TiC nanocomposite powder or mechanically-mixed nano-Ti-TiC powder. The densification of Ti-TiC nanocomposite parts was affected by the laser energy-density and the type of powder. The use of an insufficient laser energy density, of 0.25kJ/m, decreased the densification rate because of the occurrence of a balling effect. Increasing the laser energy-density, to above 0.33kJ/m, produced almost fully-dense parts. The degree of densification of ball-milled Ti-TiC nanocomposite powder was generally higher than that of directly mixed nano-Ti-TiC powder. The TiC-reinforcing phase usually had a lamellar nanostructure, with a nanoscale thickness that was entirely different to that of the initial nanoparticle morphology. The lamellar nanostructure of the TiC in ball-milled Ti-TiC nanocomposite parts could be maintained over a wide range of laser energy densities. On the other hand, the microstructures resulting from directly mixed nano-Ti-TiC powder were sensitive to selective laser melting parameters, and the TiC changed from lamellar to relatively-coarse dendritic as the laser energy density was increased. The combination of a sufficiently high densification rate and the formation of nanostructured TiC reinforcement tended to improve the tribological properties, leading to a coefficient-of-friction of 0.22 and a wear-rate of $2.8 \times 10^{-16} m^3/Nm$. Coarsening, and the resultant disappearance of nanoscale TiC reinforcement from consolidated samples made from directly mixed nano-Ti-TiC powder at a high laser energy density, considerably impaired the tribological properties.

A relatively new method for preparing metal-matrix composites containing $TiB_2$ whiskers involved binder jetting additive manufacture of a titanium matrix, periodically reinforced by the extrusion of resin containing boride particles[245]. Low-temperature pressure-free sintering then increased the strength of the green samples, and promoted chemical reaction between the matrix and the ceramic; thus resulting in the growth of boride whiskers. Studies suggested that there was a higher probability of the formation and growth of boride whiskers if the temperature in the sintering step was raised to 1400C. The Young's modulus and the yield stress ranged from 1.6 to 3.7GPa and from 83.9 to 165MPa, respectively, and the sample stiffness was considerably increased by increasing the temperature and the volume fraction. Samples which were sintered at up to 1400C exhibited a 6.4 and 15.2% improvement in stiffness, respectively, even though only small, 2% and 4%, fractions of ceramic were incorporated. There were parallel changes of 4.5 and 19% in the density of the porous matrix. It is to be noted that, although titanium-boride reinforced titanium-matrix composites have a higher wear-resistance and strength, they can also have a much poorer toughness and ductility; as compared with those of titanium and its alloys. In order to reduce these problems, a three-dimensional quasi-continuous network within the composites can be created by *in situ* laser deposition-additive manufacturing[246].

Laser engineered net shaping has been used[247] to create $Ti-Al_2O_3$ compositionally graded structures which comprised various sectors: Ti-6Al-4V, Ti-6Al-4V plus $Al_2O_3$ composite, pure $Al_2O_3$ ceramic. Each section had a unique microstructure and phase content. The pure $Al_2O_3$ sector had the greatest hardness: $2365.5HV_{0.3}$.

The effect of the powder flow-rate on the properties of titanium alloy powder, laser metal deposited onto a titanium alloy substrate was determined by varying the rate from 1.44 to 7.2g/min, while the laser-power, gas flow-rate, laser spot-size and scanning speed were fixed at values of 3kW, 2.l/min, 2mm and 0.004m/s, respectively. The surface roughness increased when the powder flow-rate was increased, and a minimum average surface roughness of 4.5μm was found for lowest powder flow-rate above[248]. Feedstock powders of commercial-purity titanium with 3 or 6wt%BN were prepared[249] by roller-milling and additive manufacturing using laser engineered net shaping. The oxidation rate was deduced from thermogravimetric analyses after 50h at 1000C. The average instantaneous parabolic constants for commercial-purity titanium, 3%BN and 6%BN were 41.2, 28.6 and 18.2mg$^2$/cm$^4$h, respectively. The titanium had an acicular α-Ti microstructure. Following thermogravimetric analysis, large equiaxed grains were observed with TiO$_2$ formation at the grain boundaries. This increased the hardness. With BN additions, plate-like TiN and needle-like TiB secondary phases also appeared. The hardness values of commercial-purity titanium, 3%BN and 6%BN were 256.9, 424.0 and 548.3HV$_{0.2}$, respectively.

Titanium-matrix composites reinforced with TiB and TiC have been produced[250] by using laser additive manufacturing, showing that the *in situ* formed reinforcement surrounded incorporated B$_4$C particles and that pro-eutectic TiB precipitated from the melt. Under the action of a high laser power penetrating into the powder-bed, the upper surface appeared to be smooth due to convection within the pool. With increasing laser power, the whisker-like TiB and granular TiC phases grew and coarsened due to the incoming energy and appreciable diffusion of boron and carbon. A microhardness of about 577.1HV$_{0.2}$ could be obtained by using a relatively low laser-power. In recent work, Ti-6Al-4V powder has been processed together with much finer titanium diboride powder, following satelliting treatment, thus producing large Ti-6Al-4V particles which are encased in finer titanium diboride[251]. The resultant composites contain boride needles and an associated increased hardness, with eutectic TiB precipitates dispersed in an α-β titanium matrix containing a few partially-melted Ti-6Al-4V and TiB$_2$ particles. Satelliting of TiB$_2$ powder onto Ti-6Al-4V particle surfaces significantly improved the homogeneity of the composite with its randomly oriented uniform distribution of TiB needles. The hardness of the composites ranged from 440 to 480HV.

There is yet another use, for laser additive manufacturing, which arises from the fact that the number of alloys which are available in the form of powder is rather small. One solution[252] is to use a mechanical mixture of elemental powders to manufacture parts via laser additive *in situ* synthesis of the desired alloy from elemental powders. This was done for the alloys, Ti-5Al, Ti-6Al-7Nb and Ti-22Al-25Nb, with titanium, aluminium and niobium particles being used as initial powders. Selective laser melting technology could thus be successfully used to synthesize α-titanium Ti-5Al alloy having an homogeneous composition and good mechanical properties. In order to dissolve fully the niobium particles, and produce an homogeneous chemical composition and microstructure in Ti-6Al-7Nb and Ti-22Al-25Nb, subsequent heat treatment was required.

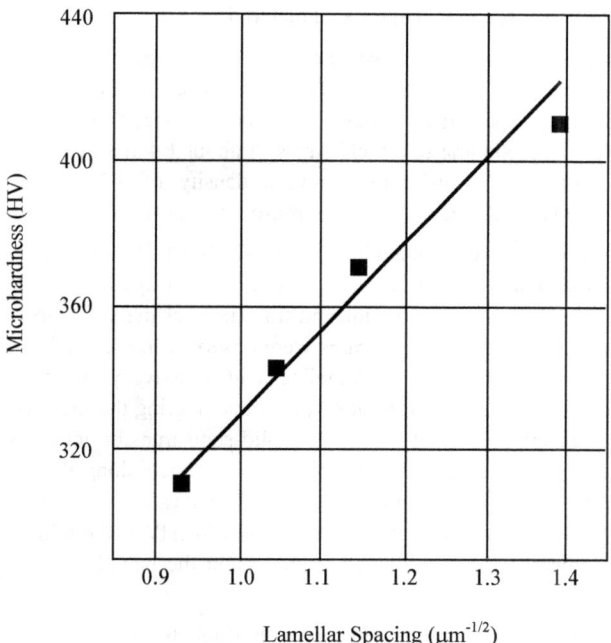

*Figure 22. Hardness of the lamellar phase in Ti-47Al-2Cr-2V*

Post-production heat treatment of additively manufactured γ-TiAl alloys, prepared via gas tungsten arc welding, promoted[253] the formation of the γ-phase throughout most of the component during holding at 1200C for 24h. A fully lamellar structure formed meanwhile in the near-substrate zone. The effect of holding at 1060C for 24h was very different, and gave a fine lamellar structure, with differing sizes in the bulk and near-substrate zones. These microstructures determined the mechanical properties of the heat-treated samples, with those held at 1200C for 24h having lower ultimate tensile strengths and microhardness values, but a greater ductility, than those of as-prepared samples

Materials Research Forum LLC
https://doi.org/10.21741/9781644900635

which had not been heat treated. Holding at 1060C for 24h led to higher ultimate tensile strengths and microhardness values, but a lower ductility. Because of the homogeneous microstructure which existed in the bulk following post-production heat treatment, the tensile properties were similar in the build and travel directions, thus erasing much of the anisotropy which is found in as-prepared material.

Electron-beam melting of precursor atomized powder produced[254] $\gamma$-TiAl intermetallic equiaxed structures with a small ($\sim$2$\mu$m) grain size and lamellar $\gamma/\alpha_2$-Ti$_3$Al colonies having an average spacing of 0.6$\mu$m. The average residual micro-indentation hardness was 4.1GPa; corresponding to a yield stress of about 1.4GPa (Tabor rule) and a specific yield stress of 0.37GPacm$^3$/g (assuming a density of 3.76g/cm$^3$). This was to be compared with the 0.27GPacm$^3$/g of electron-beam melt-fabricated Ti-6Al-4V products.

A study of laser additively manufactured Ti-47Al-2Cr-2V alloy showed[255] that the last deposited layer consisted of a fully lamellar microstructure, while equiaxed grains formed, due to the columnar-to-equiaxed transition, at the very top layer. Alternatively arranged columnar and equiaxed grains were observed in the stable zone. The columnar grain region comprised columnar lamellae, with massive $\gamma$ phase. In equiaxed grains, there was a duplex microstructure of $\alpha_2|\gamma$ lamellae. During thermal cycling, the columnar grains changed into equiaxed grains via solid-state transformation, so that alternating columnar and equiaxed grains were eventually arranged along the deposition direction. Changes in the microstructure produced corresponding variations in the microhardness, which decreased from 410HV at the top, to about 310HV at the middle, of the plate. The variation in microhardness depended mainly upon the lamellar spacing (figure 22) and the grain size.

The modulated laser powder-bed fusion method has been used[256] to prepare cylindrical samples of Ti-5553 (Ti–5Al–5Mo–5V–3wt%Cr) alloy. Low and high volumetric energy densities led to an increase in porosity, with the highest relative density (99.92%) and surface roughness of less than 12$\mu$m resulting from an energy density of 112J/mm$^3$, and also imparting a uniform hardness distribution of 295HV. The $\omega$ phase was present in samples having the highest density. High energy densities led to *in situ* precipitation-hardening, due to the nucleation of $\alpha$-Ti needles in the $\beta$-Ti phase matrix. Due to the inhomogeneous size-distribution and volume fraction of the $\alpha$-Ti needles along the build direction, a non-uniform hardness was nevertheless obtained when high volumetric energy densities were used.

Electron-beam melting was used[257] to process Ti-55511 (Ti-5Al-5Mo-5V-1Cr-1Fe), a near-$\beta$ titanium alloy. The energy input ranged from 10 to 50J/mm$^3$, the cathode current was between 4.5 and 19.5mA and the scanning speed was 1080 to 23400mm/s. Four

Additive Manufacturing of Metals

Materials Research Foundations 67 (2020)

Materials Research Forum LLC

https://doi.org/10.21741/9781644900635

types of upper surface could be discerned: flat, orange-peel, single-pores, swollen. The best results were found at an energy of 30J/mm$^3$, giving a flat surface and relative density of better than 99.9%. The scanning speed had the greatest effect: the lower the scanning speed, the higher was the aluminium loss. At low scanning-speeds, the $\alpha$-phase volume was about 78%. A higher scanning-speed resulted in that content becoming 61%. The dimensions of the lamellar, and the fraction of $\alpha$-phase, strongly affected the hardness and led to results ranging from 360 to 430HV.

An oxide-dispersion strengthened alloy, Ti-45Al-3Nb-0.2Y$_2$O$_3$at%, has been developed[258] with a view to its use in additive manufacturing. The usual dual-phase, $\alpha_2$-Ti$_3$Al and $\gamma$-TiAl, duplex and near-lamellar microstructures were found in as-processed material. The oxide particles were homogeneously distributed throughout the matrix after processing in the liquid state. The hardness of the consolidated material was higher for oxide-containing samples. In further work[259], the effect of the Y$_2$O$_3$ upon the mechanical properties and oxidation resistance was studied; especially the mechanical properties at 293 to 1073K and the oxidation resistance at 1073K of spark plasma sintered and direct metal deposited samples. At room temperature, a 34% higher yield stress and a 14% higher ultimate tensile strength, but a 17% lower ductility, were found for oxide dispersion strengthened, as compared with non-strengthened, material. The strengthened form also exhibited better strength retention at up to 1073K. The strengthened direct metal deposited material exhibited a similar deformation behaviour to that of sintered material, but was prone to early fracture due to the presence of residual porosity. The addition of Y$_2$O$_3$ increased the oxidation resistance of both sintered and direct metal deposited material: the parabolic growth constants decreased by 49% and 75% for sintered and direct metal deposited material, respectively. In sintered samples, the dispersoid size changed slightly, from 29 to 26nm after 987h at 923K and to 32nm after 924h at 1073K, thus illustrating the high stability of the particles. Widespread grain-boundary pinning by the particles then contributed to good microstructural stability.

A study[260] of the effect of heat treatment upon the laser additive manufacturing of Ti-5Al-2Sn-2Zr-4Mo-4Cr (TC17) showed that the as-deposited sample contained traces of a prior mixed $\beta$-phase grain-structure comprising equiaxed and columnar grains, intragranular ultra-fine $\alpha$-phase laths and widespread continuous grain-boundary $\alpha$-phase. Following pre-annealing in the $\alpha$-$\beta$ region (840C), and standard solution and aging treatments, the continuous grain-boundary $\alpha$-phase was coarser and the nearby precipitate-free zone had become a zone filled with ultra-fine secondary $\alpha$-phase but no primary $\alpha$-phase. Following pre-annealing in the $\beta$ region (910C), all of the $\alpha$-phases had transformed into $\beta$ phase and the alloying elements were uniformly distributed near to the grain boundary. Discontinuous grain-boundary $\alpha$-phase, and a uniform mixture of the other forms of α-phase near to the grain boundary were present after subsequent solution and aging treatments. These heat treatments could improve the tensile properties.

The creation of parts by means of laser additive manufacturing, and their joining by the use of electron beam welding is a feasible means of manufacturing large constructions having reduced internal stresses. As an example, 2 laser additively manufactured Ti–6.5Al–3.5Mo–1.5Zr–0.3Si plates were welded[261] without introducing defects. The microstructure of the base metal had a typical basket-weave morphology, with lamellar α-phase in a β-phase matrix. In the heat-affected zone, some of the primary α-phase transformed into β-phase while some very fine lamellar α-phase precipitated out. Due to the rapid solidification, a large number of acicular α′-phase particles formed in the fusion zone, producing the maximum microhardness. All of the tensile samples failed in the base-metal region, with an intergranular dimpled fracture surface.

The direct energy deposition laser additive manufacturing process was used[262] to deposit β-type Ti-15Mo biomedical alloy, a particular problem being the formation of a strong metallurgical bond with minimum track dilution. Statistical analysis of the results showed that the minimum degree of track dilution was achieved by using the maximum laser-power and the minimum scan-speed and powder feed-rate. By using these optimum parameter-settings, a microhardness of 385HV and an elastic modulus of 73GPa were obtained.

Laser engineered net shaping was used[263] to create Ti-Si-N coatings, having 3 different Ti/Si ratios, on a commercial-purity titanium substrate. The coatings had a gradated microstructure with *in situ* formed phases. High hardness values were found for all of the upper surfaces, in the order: $2093.67HV_{0.2}$ for Ti-10%Si-N, $1846HV_{0.2}$ for 100%Ti-N and $1375.3HV_{0.2}$ for Ti-25%Si-N. The wear resistance, on the other hand, was more dependent upon the silicon content, and samples with a higher content had a better wear resistance.

Selective laser melting has been used[264] to form titanium–tantalum mixed ball-milled powders having tantalum contents ranging from 0 to 25wt%. With increasing tantalum content, the microstructure of the resultant products changed from lath α grains to acicular α′ plus primary cellular β grains, together with a gradual suppression of the martensite transformation. The β-stabilizing effect of tantalum encouraged the formation of a β (Ti,Ta) solid solution phase. The increasing tantalum content increased the tensile strength from 641 to 1186MPa and the microhardness from 257 to 353HV, due to the combined effect of Hall-Petch strengthening and solid-solution strengthening. The Young's modulus decreased from 115 to 89GPa however, due to the increasing fraction of β phase. The corrosion resistance of the alloys was also improved, with little surface pitting, due to an increasing amount of $Ta_2O_5$.

*Ti-6Al-4V*

As is usual in engineering and metallurgy, the experience gathered in one domain can spill over into others. In the case of titanium-based alloys, for instance, it so happens that their properties make them compatible with human tissue and the very same alloy which is valued by the aerospace industry may be of use to the medical profession. One particular alloy of which this is true is Ti-6Al-4V. This alloy has essentially become a benchmark by which to gauge the capabilities of each metal additive manufacturing process as it is developed. It will be informative therefore to consider this alloy in some detail and also introduce in passing some of the standard techniques used in the additive manufacture of metallic components. As hinted above, shaped metal deposition offers great advantages in the case of this alloy. The resultant components exhibit large columnar prior-β grains having a Widmanstätten α/β microstructure, with the β-grains being slightly tilted in the direction imposed by the temperature-field of the moving welding torch[265]. The ultimate tensile strength is typically between 929 and 1014MPa, depending upon the location and orientation of the test samples. When the samples are pulled in the direction perpendicular to the deposited layer, the strain to failure is 16%. When they are pulled in a direction parallel to the deposit, the strain to failure is about 9%.

The most common current types of metal additive manufacturing are selective laser sintering, selective laser melting, direct metal laser sintering, electron-beam melting and laser engineered net shaping. All of these are of interest for the manufacture of Ti-6Al-4V-based biomedical implants[266]. Additive manufacturing was adopted by the medical profession in the late 1980's because it made it possible to obtain solid prostheses which simplified surgery, and reduced risk. In this connection, samples of Ti-6Al-4V were produced[267] by direct metal laser sintering, yielding a microstructure consisting of hexagonal martensite with an acicular morphology. This was associated with an average microhardness of 370HV, an ultimate tensile strength of 1172MPa, a yield strength of 957MPa and an elongation-to-fracture of 11%. Because melting takes place in a vacuum environment, electron-beam methods can produce parts having higher densities than in the case of selective laser melting, especially when titanium alloys are involved. Electron-beam additive manufacturing involves layers of deposited metal and can exhibit appreciable variability. One important variable is the deposit bead-width, as it affects both the build geometry and the microstructure. The optimum bead-width is between 1.0 and 1.3cm, but in practice it tends to range from 0.8 to 2.0cm[268]. As in the case of other alloys, it has been proposed that additive laser deposition and machining could be used to repair titanium parts. An early study[269] already examined questions of preparing the damage site, laser deposition, machining, sample preparation and mechanical testing. Practical tests showed that repaired part could actually be stronger; the key requisite of all repairs.

The strain-rate effect upon specimens prepared by using electron-beam methods was investigated[270] by using rates which ranged from $10^{-2}$ to $10^{-4}$/s. This showed that in-plane values of the elastic modulus, yield strength and ultimate tensile strength were considerably higher than the out-of-plane values. This was attributed to the presence of defects between the layers or to imperfect bonding of the layers. A strong positive strain-rate sensitivity was evident in the out-of-plane direction, while there was no strain-rate dependence of the in-plane directions.

The surface modification of Ti-6Al-4V alloy has been achieved[271] by means of additive manufacturing, using a ball-milled mixture of titanium and SiC powders. Laser cladding layers of Ti-10%SiC and Ti-20%SiC were deposited onto the Ti-6Al-4V substrate by using a single-channel feeding system. Use of this system with the ball-milled powder mixture guaranteed a more homogeneous SiC distribution and hence better fluidity during the laser cladding process. Chemical reaction occurred between the titanium and the SiC particles in the composite layers, with the reaction products being TiC and $Ti_5Si_3$. The average three-point bend strength of the sample with the Ti-20%SiC coating was 515MPa at room temperature. This cladding layer increased the hardness of the Ti-6Al-4V matrix from 339.1 to 932.2 HV, and markedly improved the wear resistance.

Wear-resistant TiB-TiN reinforced Ti-6Al-4V composite coatings have been deposited[272] onto titanium substrates by using laser-based additive manufacturing technology. The Ti-6Al-4V powder, pre-mixed with 5 or 15wt% of boron nitride powder, was used to create TiB-TiN reinforcement *in situ* during laser deposition. The high temperatures which were generated led to the formation of TiB and TiN. With increasing BN content, from 5 to 15wt%, the Young's modulus of the composite coating increased from 170 to 204GPa. There was a marked increase in the wear resistance with increasing BN concentration. For given test conditions, TiB-TiN composite coatings with 15wt%BN exhibited an order-of-magnitude lower wear-rate than that of CoCrMo alloy, a common orthopaedic implant material. The average surface hardness of the composite coatings increased from 543 to 877HV with increasing BN concentration. The coatings were non-toxic, and had a similar biocompatibility to that of commercial-purity titanium.

Selective laser melting was used to prepare Ti-6Al-4V which was reinforced with 1% or 2.5wt% of nano-sized yttria-stabilized zirconia[273]. The material had a very high density, with a near-alpha microstructure. No change in grain-shape was observed following the addition of the zirconia, and columnar and lamellar grains remained. The microhardness of the alloy was markedly increased, from 340 to 511HV, due to the zirconia addition, with the maximum yield strength and compressive strength being 1302 and 1751MPa, respectively, as compared with the values of 840 and 1250MPa, respectively, for plain selective laser melted Ti-6Al-4V.

Laser engineered net shaping was used[274] to process a mixed coating of Ti-6Al-4V powder and calcium phosphate in an oxygen-free nitrogen-argon atmosphere. The resultant coatings were composites of titanium nitride and calcium titanate in an α-Ti matrix. The hardness was increased by some 148%, to 868HV, as compared with that of the untreated Ti-6Al-4V substrate. When the tribological properties, against alumina, were determined in de-ionized water, the wear damage was reduced by about 91%, as compared with the untreated Ti-6Al-4V substrate. The latter substrate released some 12.45ppm of titanium ions during wear, while the Ti-6Al-4V and 5% calcium phosphate coating, processed in an argon-nitrogen atmosphere, released only some 3.17ppm of ions under equivalent testing conditions. The overall coefficient-of-friction also decreased due to the addition of calcium phosphate.

A comparison was made[275] of the selective laser melting and electron beam melting methods, as used for the fabrication of titanium implants. When selective laser melting was performed under 0.4 to 0.6vol% of oxygen in order to improve the mechanical properties of Ti-6Al-4V alloy, the latter exhibited a high anisotropy of the mechanical properties and a better (1246 to 1421MPa) ultimate tensile strength as compared with that (972 to 976MPa) of electron beam melted material or that (933 to 942MPa) of wrought material. The microstructure and phase composition depended upon which method was used. Additive manufacturing led to the formation of long epitaxial grains of prior β-phase. Equilibrium α-β and non-equilibrium a'-martensite were found for electron-beam and laser processing, respectively. Although heat-transfer during the layer-by-layer deposition caused changes in the aluminium content, neither method produced any cytotoxic effects.

In this connection, X-ray computed tomography studies have been made[276] of calcium phosphate and polycaprolactone coatings on the Ti-6Al-4V scaffolds which are used as implants. A cylindrical scaffold had a greater porosity on its inner side than on its outer side, thus neatly mimicking trabecular and cortical bone, respectively. Prismatic scaffolds had a uniform porosity. The scaffold surfaces were modified, by dip-coating with phosphate or polycaprolactone, in that this improved the biocompatibility and mechanical properties. X-ray and synchrotron-radiation computed tomography data revealed structural and morphological defects in the coatings, with small platelet-like and spiderweb-like feature being observed in calcium phosphate and polycaprolactone coatings, respectively.

In laser-based direct-energy deposition, processing can be controlled by using a closed-loop system in which thermal sensing of the melt pool surface is used to adjust the laser settings so as to maintain a constant surface geometry. A heat-transfer and fluid-flow laser-welding model have been used[277] to examine how changes in the processing parameters affect the relationship between the upper and sub-surface temperatures and the solidification behaviour of Ti-6Al-4V. Numerical simulation showed that liquid pools having similar upper-surface geometries could nevertheless involve very different penetration depths and volumes. The melt-pool surface area was in fact a poor indicator of the cooling-rate at various locations in the melt pool and could not be used to estimate the final microstructure or consequent mechanical properties. As the build-temperature increased and the power-level was adjusted in order to maintain a constant surface geometry, variations could occur in the solidification behaviour and greatly affect the final microstructure.

A new heat-source model has recently been proposed[278] for simulating the temperature-rise occurring during laser deposition melting additive manufacturing. The temperature histories found for various layers can be used to predict microstructural changes at a number of scales. This shows that re-melting and re-heating, due to the differing temperature histories of the layers, are the main factors affecting microstructural changes during laser-based additive manufacturing. The flow stresses at various strain-rates and temperatures can then be predicted. In particular, the flow stress is higher in the middle of the as-built product and lower near to the top or bottom surfaces. An increase in the laser scanning-rate can lead to an increase in the size of the heat-affected zone. This thus explains the formation of larger grain sizes and more $\alpha$-phase production at higher scanning-rates, and a higher $\alpha$-phase volume fraction is in turn the cause of improved mechanical properties at higher scanning-rates.

The thermal conductivities of metal powders (Ti-6Al-4V, Inconel-718, Inconel-625, 17-4, AISI316L) commonly used for powder-bed additive manufacture were measured[279] using the transient hot wire method, at 295 to 470K, under argon, nitrogen or helium gas pressures of 1.4 to 101kPa. The results indicated that the gas pressure and composition had a marked effect upon the effective thermal conductivity of the powder, whereas the metal-powder properties and temperature did not.

Microstructure development during additive manufacturing is typified by columnar grains which are produced in the heat-affected zone during solidification of the melt pool. A study[280] of the heat-affected zone which was produced during the laser metal deposition of grade-5 titanium alloy, with the laser power being varied from 0.8 to 3.0kW while maintaining other process-parameters constant, showed unsurprisingly that the length of the heat-affected zone increased with increasing laser power.

Based upon the Rosenthal solution for a moving-point heat-source, simulation-based processing maps were developed[281] for the thermal conditions which control microstructural features such as grain- size and grain-morphology, during beam-based deposition in semi-infinite geometries where a steady-state melt pool exists far from the free edges. The Rosenthal solution was modified, so as to include the effects of those free edges, by superposing two point-like heat-sources which approached one another; the line-of-symmetry representing the free edge. This resulted in an exact solution for the case of temperature-independent properties. The dimensionless results for the melt-pool geometry were then plotted as a function of the distance from the free edge and these solidification maps predicted trends in the microstructure of Ti-6Al-4V. One notable result was that the melt-pool geometry appeared to be more sensitive to the free edges than to the solidification microstructure.

The effects of many of the process variables and alloy properties upon the structure and behaviour of additively manufactured parts can be characterized[282] by using just 4 dimensionless numbers, including those of Peclet, Marangoni and Fourier. This applies to the structures and properties of components made from Ti-6Al-4V, AISI316 stainless steel and Inconel-718 powder. The temperature fields, cooling rates, solidification parameters and thermal strains were predicted by using three-dimensional transient heat-transfer and fluid-flow models. The results showed that fusion defects in fabricated components could be affected by increasing interlayer bonding using high levels of dimensionless heat input. The formation of deleterious intermetallics, such as Laves phases in Inconel-718, could be suppressed by means of a low heat-input. This resulted in a small molten pool, a steep temperature-gradient and a rapid cooling-rate. An improved interlayer bonding could be ensured by imposing high Marangoni numbers, which resulted in a vigorous circulation of the liquid metal, large pool-sizes and greater penetration-depth. Choice of a high Fourier number imposed rapid cooling, low thermal distortion and a high ratio of temperature gradient to solidification growth rate; leading of course to a greater tendency to plane-front solidification[283].

In a potential medical application[284], the use of 170W laser sintering created fully-dense specimens, while other specimens consisted of a fully-dense outer skin and a partially-sintered porous inner core. The outer skin was scanned using 170W, thus producing thicknesses of 0.35, 1.00 or 1.50mm. The inner core was scanned using 43 or 85W. Slightly greater porosity was associated with lower laser powers, but the degree of porosity in the core was not related only to the laser power. That is, thinner skins resulted in a higher porosity of the core for a given power. The lowest Young's modulus which was measured was 35GPa, and this was close to that of bone. It was associated with a laser power of 43W and a skin thickness of 0.35mm, with the core making up 74% of the total volume. In the biomedical field, additive manufacturing produces near-final shape products from titanium alloys, but finishing operations may be required in order to obtain the desired geometrical tolerances. A comparison[285], with wrought commercial alloy, of tool crater-wear occurring during the turning of Ti-6Al-4V components produced by electron beam melting or direct metal laser sintering, showed that there was a correlation between the mechanical and thermal properties of the alloy, and the crater wear which occurred. Liquid nitrogen was used as a coolant in order to reduce crater wear. Changes in porosity during additive manufacturing may be an indication of problem-points within the process. Some work has been done[286] on monitoring porosity during additive manufacture.

The application of additive manufacturing to biomedicine offers great promise with regard to optimum geometry, reduced risk of rejection and improved ergonomics. Components made from Ti-6Al-4V, and produced by direct metal laser sintering, unfortunately require further heat treatment in order to impart optimum mechanical properties and relieve internal stresses. Direct metal laser sintered samples exhibited[287] various microstructures following heat treatments at 650 to 1050C, the main differences involving the nucleation and growth of β-phase which could have impair the corrosion behavior. The latter was investigated by means of cyclic potentiodynamic polarization in phosphate-buffered saline solution at 37C. During anodization, an oxide film grew on the surface which consisted of a barrier plus a nanoporous layer which was doped with fluorine ions. The passive current density was reduced by some 2 orders of magnitude under all conditions, regardless of morphological differences between the layers. This not only improved the corrosion resistance, but also decreased any potential ion release into a bloodstream.

Selective laser melting has been used, for example, to make a hip implant from plasma-atomized Ti-6Al-4V powder. Computer-tomography data on a deformed bone structure was used to print a hip bone model, 3-dimensionally, from polyamide. Then an implant prototype was made from polymer while accounting for particular anatomical variations. The implant model was then scanned 3-dimensionally in order to obtain a computer-aided design file on the implant, and this was further improved by partially texturing the surface. The metal implant was then finally produced by selective laser melting, and annealed so as to impart a better combination, of tensile strength and elongation, via partial decomposition of the martensitic phase. Such an application of additive manufacturing to the production of a hip implant decreased the operating time and lessened the risk of infection.

A wire-feed pulsed plasma arc method was used[288] to produce thin walls, with the heat input being gradually decreased layer-by-layer. The thin wall comprised various morphologies, including the epitaxial growth of prior β-grains, martensite and horizontally layered bands of Widmanstätten phase, and their incidence depended upon the heat input, thermal cycles and gradual cooling-rate during deposition. A reduction in the heat input of each bead and the use of a pulsed current refined the microstructure and strengthened the thin wall. The average yield strength and ultimate tensile strength attained 909 and 988MPa, respectively, while the elongation reached some 7.5%.

Laser metal deposition has in fact long been used as a coating technology, but the present-day ability to carry out 3-dimensional deposition led to its new use in additive manufacturing. In the laser metal deposition of Ti-6Al-4V, only a coaxial laser metal deposition powder nozzle is used to create the shielding-gas atmosphere, thus permitting a high geometrical flexibility. Products having a high aspect-ratio, involving hundreds of layers, can be manufactured. In one demonstration[289], cylindrical specimens were first manufactured using a standard shell-core build-up approach, and the fracture properties were determined. Using the results, a modified approach was used which incorporated variable track overlap ratios in order to achieve constant growth in the shell and core area. Cylinders consisting of more than 240 layers, and having a height of more than 120mm were manufactured.

A simplified numerical model has been used[290] to predict layer-heights and temperature distributions during laser metal deposition of titanium alloy. An arbitrary Lagrangian Eulerian free-surface motion was made to be directly dependent upon the powder feeding-rate. In the case of thin walls of Ti-6Al-4V, the numerical results compared well with experimental data on the geometrical features, molten-pool sizes and temperatures. Although the model did not take account of coupled hydraulic-thermal effects, it provided a more realistic local geometrical description of additive layer manufacture than did simpler thermal models, as well as offering much shorter calculation-times when compared with sophisticated approaches that considered thermocapillary fluid-flow. Equiaxed or columnar microstructures were predicted for Ti-6Al-4V walls by using available microstructural maps together with the local thermal gradients and solidification rates which were furnished by finite-element calculations for points near to the solidification front.

Direct metal laser sintering is capable of producing complex geometries, and samples of Ti-6Al–4V were prepared by direct metal laser sintering and stress-relief heat-treatment

(650C, 3h), following processing at 850, 950 and 1050C for 1h and furnace cooling. The higher the heat-treatment temperature, the higher was the ductility and the lower was the mechanical strength. This was attributed[291] to the nucleation and growth of α and β phases. The ductility in compression was higher than that in tension, due to a difference in the behaviours under tensile and compression loading. Samples which were treated at 950 and 1050C were considered to offer the best combination of mechanical properties for use in implants. The static and dynamic compressive properties of pure titanium structures were compared[292] with the results reported for identical structures made from Ti-6Al-4V or tantalum. This showed that porous Ti-6Al-4V was then still the strongest material for use in static load applications. Pure titanium exhibited a mechanical behavior which was similar to that of tantalum, and was then still preferred material for use as cyclically loaded porous implants.

Customized implants and reconstruction plates are currently manufactured using direct metal laser sintering or electron beam melting. Alloy particles and surface flaws tend to exist on the as-processed surfaces, and these are removed by polishing. The surface roughness of electron-beam processed alloy was found to be higher than that of direct metal laser sintered alloy. The cytotoxicity of the samples was compared, showing that the mean cell viability on all of the samples was 82.6%, and it was concluded[293] that the specimens were non-cytotoxic in both the as-processed and polished conditions.

Laser-based additive manufacture of metals from powder can be achieved in two main ways: directed-energy deposition and powder-bed fusion. Metallic powder is fed to a location and melted locally using the laser heat source. During deposition, the material experiences rapid cooling and solidification and, with the addition of subsequent layers, the material within the component undergoes rapid thermal cycling. Transient fluid dynamics modelling of the Ti-6Al-4V melt pool can be used to characterize the relevant process parameters, such as melt-pool geometry, beam power, beam speed, beam diameter, and temperature profile. It is known that the cooling-rate in the melt pool is directly related to the part's microstructure and thus greatly influences the strength and hardness. The cooling rate is in turn related to the basic heat-transport process, which involves a moving heat-source and rapid self-cooling.

The microstructure and indentation hardness of Ti-6Al-4V components which had been prepared by pulsed or continuous-wave laser beam processing were studied[294], revealing that pulsed-beam build-up led to no significant variations in lath-width or indentation hardness with increasing build-height. On the other hand, material which had been deposited using a continuous-wave beam exhibited a statistically significant decrease in hardness and an increase in lath-width near to the middle of the built-up sample. The lower variability in the case of pulsed-beam treatment was attributed to the occurrence of rapid cooling within the sample.

Additively manufactured Ti-6Al-4V was subjected to rotating bending fatigue tests at room temperature and 250C[295]. These specimens had an acicular α-β microstructure whereas the conventionally manufactured alloy consisted of α-phase in a β-phase matrix. Round and crevice-like defects with sizes of up to 50μm were found in the additively manufactured alloy, but not in the conventionally manufactured samples. The fatigue strengths, at $10^7$ cycles, of conventional material were 625MPa at room temperature and 475MPa at 250C, while those of the additively manufactured alloy were 300MPa at room temperature and 250MPa at the higher temperature. The fatigue cracks in the additively manufactured specimens began from defects at both temperatures.

The related alloy, Ti-6Al-7Nb, has also been found to be useful in the field of osteopathic and dental prosthetics, where a high surface quality is required. Bone scaffolds for tissue-engineering require the use of cellular structures which have pore diameters that encourage the growth of osteoblasts. In a study[296] of the quality of complex components made by powder-bed selective laser melting, test-pieces were subjected to chemical polishing in order to improve the surface quality and to remove loose particles which were trapped within the porous structure. The resultant surface roughness, and the reduction in the number of un-melted powder particles on the scaffold surface, were governed mainly by the chemical composition and by the bath concentration, together with the means of medium delivery and exchange during the process.

The quality (packing density, surface uniformity) of the powder layer is of course a critical factor affecting any components produced using powder-bed metal additive manufacturing; regardless of whether that involves selective laser melting, electron beam melting or another technique. A computational model has been used[297] to study the effect of powder cohesion upon powder re-coating during additive manufacturing. It was based upon a discrete-element analysis which took account of particle-particle and particle-wall interactions via frictional contact, rolling resistance and cohesive forces. The proposed adhesion force-law was couched in terms of the pull-off force, resulting from the surface energy of powder-particles, together with a van-der-Waals force curve. Spatial mean values and standard deviations of the packing fraction and surface profile were used to characterize the powder-layer quality. As an example, the size-dependent behavior of plasma-atomized Ti–6Al–4V powders during re-coating process was considered. The model was used to predict the angle-of-repose of spherical Ti-6Al-4V powders, and the surface energy implicated in the adhesion force-law was deduced by fitting the angle-of-repose to the results of numerical and experimental funnel-tests[298]. This yielded an effective surface energy of 0.1mJ/m$^2$; a value which was considerably lower than the typical experimental values of 30 to 50mJ/m$^2$ for flat metal contact surfaces. A decreased particle size and increased cohesion led to an appreciably impaired powder-layer quality in terms of low, highly-variable packing-fractions and very non-uniform surface profiles. In the case of powders having a median particle-diameter of 17μm, cohesive forces predominated over gravity forces by 2 orders of magnitude. This led to low-quality powder layers which were unsuitable for laser melting without then adding further finishing steps. It was concluded that the neglect of cohesive forces led to serious underestimation of the angle-of-repose, and thus to an insufficient appreciation of bulk-powder behavior

Also related to the angle-of-repose of a powder and its rheological behavior is its flow or shear motion and thus its ability to spread, as a uniform layer of loose powder having a suitable thickness, over a specified area. A rheometer which is able to assess the bulk flow performance has been used to characterize metal additive manufacturing powder. When combined with powder-dynamics modelling, the rheometer provides data which quantify the degree of so-called spreadability. That is, the ease with which a powder spreads under given conditions. Powder-dynamics modelling using discrete-element methods can simulate powder behaviours such as segregation, porosity and surface roughness, but should also allow for true particle shapes and sizes. This can be too computationally expensive. A simpler method[299] is to compare a virtual sample, comprising monodispersed spherical particles, with angle-of-repose and rheological data for a real sample comprising almost-spherical particles of comparable size. As an example, results for a sample consisting of 2mm glass beads were compared with those for Ti-6Al-4V powder with a particle-size of 100 to 250μm and computed data from a model involving 1,300,000 particles of 250μm Ti-6Al-4V powder.

As an example of the value of additive manufacture, a nanopositioning flexing device made from Ti-6Al-4V was created by electron-beam melting[300]. This alloy offers a high biocompatibility, corrosion resistance and strength, but is difficult to machine by conventional means. Additive manufacturing permitted the construction of a 3-dimensional nanopositioning device which permitted considerable mechanical displacement within a compact structure. Comparison of the mechanical properties of electron-beam printed Ti-6Al-4V cantilevers, with those of bulk metal cantilevers, showed that - due to the porous surfaces - the printed cantilevers behaved like a softer material having an average Young's modulus of 41GPa. This calculation considered only the outer dimensions. Upon introducing inner widths of 0.51 to 0.53mm into the beam-width of 0.7mm, the resultant Young's modulus of 90 to 120GPa was comparable to the 108 to 120GPa which was reported for bulk Ti-6Al-4V. Mechanical levers which were printed within the device amplified the output of a piezoelectric actuator by a factor of 6, and thereby displaced a positioning platform which was supported by a network of parallel supporting beams. The maximum displacement (47.4µm) occurred at a driving voltage (150V) and the resonant frequencies measured for the x- and y-axes were 1854 and 1858Hz, respectively. For triangular sweeps of 16 and 122Hz, the positioning errors were less than 200 and 500nm and the time-delays were 0.85 and 2.48ms, respectively.

Powder-bed electron-beam additive manufacturing is a relatively new technique in which the metal powder is melted under vacuum by using a high-energy heat source. Electron-beam melting in general has been applied to hexagonal close-packed Ti-6Al-4V, nickel-based face-centered-cubic René-142 superalloy and body-centered cubic pure iron, covering an overall melting-point range of 1375 to 1630C. The process relies upon the availability of pre-alloyed precursor powders which are selectively melted layer-by-layer.

In essence, powder-bed electron-beam additive manufacturing involves a rapid solidification process and the properties of the product depend upon the solidification behavior as well as upon the microstructure. Phase-field modelling has been used[301] to study microstructure development and solute concentrations in Ti-6Al-4V during electron beam additive manufacturing. As expected, a greater undercooling led to faster dendrite growth. The microstructure simulations predicted multiple columnar-grain growth, consistent with experimental results. One advantage of powder-bed metal processing is that the powder can be reused, and this clearly affects the cost of additively manufactured parts; especially in the case of titanium alloys. A study[302] of the effect of powder re-use upon Ti-6Al-4V powder composition, particle-size distribution, density, flow and particle morphology showed that re-using the same sample of powder 21 times for selective electron beam melting led to the oxygen content continually increasing with the number of re-uses while the aluminium and vanadium contents remained fairly stable. The powder meanwhile became less spherical with increasing re-use, and some particles already exhibited distortion and surface-roughening after being re-used just 16 times. The particle-size distribution narrowed, and few satellite particles were observed, after re-using 11 times. On the other hand, re-used powder had an improved flowability, and there was no detectable impairment of the additive manufacturing process, with the samples exhibiting very consistent tensile properties; irrespective of their location in the powder bed. Other studies[303] indicated that the microstructure evolved from a non-equilibrium to an equilibrium α and β phase state with increasing re-use. The microhardness and Young's modulus of re-used powder improved with increasing re-use. The tensile properties of bulk products which were made from mixed powders were almost equal to, or better than, those of products made from 100% fresh powder. When Ti-6Al-4V powder was recycled 30 times in an electron-beam melting system[304], the recycled powder again had a 35% higher oxygen content and the particles had a more irregular morphology, with a narrower particle-size distribution and a considerably more variable microstructure than that of virgin powder. The microstructure of the recycled powder here varied from a martensitic α′ structure, identical to that of virgin powder, to a two-phase α-β structure. This was attributed to the complex thermal history of the non-melted powder. Regardless of the differences, the particles all had roughly the same surface oxide thickness, with the excess oxygen of the recycled powder being present in the β phase. Although the use of powder in additive manufacturing is quite common, the production of powder and the choice of powder properties which give the required results, has posed some considerable problems[305]. The properties sought in the high-quality titanium metal powders used for additive manufacturing include a combination of high sphericity, density and flowability.

The sensitivity of the process to operating characteristics such as powder variations, is a major cause of scatter in the properties of electron beam additively manufactured components. A study[306] of experimental techniques and temperature measurements investigated the effects of process parameters upon the thermal characteristics of Ti-6Al-4V powder in terms of a so-called speed function index which controlled beam speed and beam current during component build-up. Electron-beam additively manufactured parts were made using speed function indices which ranged from 20 to 65, while processing temperatures were monitored via near-infrared thermography. It was found that the speed function index noticeably affected the thermal characteristics during electron beam additive manufacturing, with the melt pool-lengths being 1.72 and 1.26mm for speed function indices of 20 and 65, respectively, at a build-height of 24.43mm. The speed function index also markedly affected component quality with regard to surface morphology, surface roughness and microstructure. A higher speed function index tended to produce products having rougher surfaces, more pores and large β-grain columnar widths. Increasing the beam speed reduced the peak temperature and reduced the melt pool-size. The higher the beam current, the higher was the peak processing temperature and the larger the melt pool. Monotonically increasing the beam diameter decreased the peak temperature and the melt pool-length.

Under high-productivity conditions, the deposit temperature during electron beam additive manufacturing increases to such an extent that it is no longer possible to develop a microstructure which possesses the required mechanical properties. In one study[307], electron beam welding was used to produce deposits of Ti-6Al-4V under various conditions. A simple analytical heat-flow model of the deposit build-up was then based upon experimental deposit-temperature data. This showed that it was possible to improve the heat flow from the deposit, control the microstructure and simultaneously maintain a good productivity.

The application of non-equilibrium equations to the modelling of phase formation and dissolution during additive manufacture of Ti-6Al-4V components[308] permitted the prediction of microstructural evolution. When applied to the cases of electron-beam melting and selective laser melting, it could explain the significantly different microstructures which resulted from using those processes.

The tensile and fatigue properties of additively manufactured Ti-6Al-4V components having machined or polished surfaces can be comparable to, or better than, those of the alloy in a conventional mill-annealed condition. On the other hand, the properties can exhibit a great amount of scatter and can also be anisotropic. Post-manufacture treatments are therefore required in order to obtain the desired properties, and consistent behaviour.

Wire-based laser metal deposition is a further new method which permits the additive manufacture of complex titanium-alloy components. The examination of specimens which had been prepared by using a 4.5kW diode laser cladding system[309] showed that the mechanical properties were affected mainly by the crystal structure and, by implication, the thermal history of the sample. The high tendency of this alloy to oxidize and distort, together with its two-phase microstructure, make it a challenge to additive manufacturing ingenuity. By using a local multi-nozzle with shielding-gas technique, the intrusion of oxidation into the process could be avoided. Because a single bead was involved, the degree of distortion was also minimized. The third problem, of a two-phase microstructure, could not be avoided by process-design alone and an α-martensite structure prevailed. Post-production heat-treatment was therefore necessary.

Wire arc additive manufacturing is based upon welding technology, is eminently affordable, and permits a very high deposition rate, while Ti-6Al-4V components can be more efficiently produced by wire arc additive manufacturing than by conventional machining. The deposition of large (>10kg) components made from titanium, aluminium and steel is possible by using wire arc additive manufacturing. High deposition-rates, low material and equipment costs and a good structural integrity make the method a candidate when replacing manufacture from billets or forgings; particularly with regard to low- or medium-complexity components.

As compared with laser powder-bed methods, arc-based additive manufacturing offers the advantages of an essentially unlimited assembly space and a higher deposition rate. Its disadvantages include the limited availability of different types of wire, a wire feed-rate which is directly coupled to the heat input and an inability to create *in situ* multi-material structures. A new, so-called 3-dimensional plasma metal deposition method[310], is based upon plasma powder deposition and permits up to 4 powders to be mixed within a single layer. These powders can moreover differ with regard to material-type and proportion. This in turn permits the tailoring of local properties such as wear resistance to match expected load types and levels.

A comparison was made[311] of bulk specimens prepared by using laser deposition additive manufacturing and wire and arc additive manufacturing, using titanium alloy powder or wire. The macrostructure of specimens consisted of coarse β-phase columnar crystals, while the microstructure was of fine basket-weave type. Annealing could cause the strength to decrease slightly and the plasticity to improve. The method of additive manufacture caused little difference, but the length/width ratio of the α-phase was relatively large and the porosity was slightly higher in the wire arc additively manufactured samples in the as-prepared and annealed states, with the corresponding plasticity being 38 and 31% lower than that of laser-deposited material, respectively. The tensile fracture modes of the specimens were all the same, and were ductile.

A study was made[312] of various post-processing treatments and of their effects upon the microstructures and tensile properties of Ti-6Al-4V components produced by wire arc additive manufacturing. The relatively slow (10 to 20K/s) cooling rates occurring during β-α transformation produced Widmanstätten α-phase and permitted a balance to be made between strength and ductility. Hot isostatic pressing removed gas-porosity but did not improve the strength or ductility. Residual tensile stresses in as-prepared components markedly harm the ductility and have to be removed.

Wires made from Ti-6Al-4V have been melted using an electric arc and deposited layer-by-layer[313]. The deposits in the first and second layers were in the form of columnar crystals while the rest were equiaxed crystals. The heat generated by the arc led to good metallurgical bonding between the deposition and fusion zones and eliminated martensite. The product had a stable α-β lamellar structure and the various zones had a similar microhardness. When compared with as-cast Ti-6Al-4V, the present material contained finer initial β-titanium grains and a smaller α-β lamellar spacing as a result of the wire arc additive manufacture. The ultimate tensile strength and elongation were increased by 3.6 and 37%, respectively. The tensile fracture morphology consisted of ductile dimples, very unlike the torn-edge quasi-cleavage morphology of as-cast alloy.

The complicated heat-flow behaviour which occurs during processing plays an important role in the formation and properties of Ti-6Al-4V components which are prepared by means of wire arc additive manufacturing. *In situ* measurements have been made[314] of heat accumulation and the thermal behaviour during gas tungsten wire arc additive manufacturing. These revealed that, due to the effects of heat accumulation, the layer surface oxidation, microstructural evolution and grain size varied along the building direction of the as-prepared material. This then led to variations in the mechanical properties and fracture behaviour. It was necessary to keep the process interpass temperature below 200C in order to assure the quality of Ti-6Al-4V components which were prepared by using only localized gas-shielding.

Experimental and modelling studies have been made[315] of wire-fed electron-beam 3-dimensional printing in order to clarify the metal-transfer mechanisms, particularly with regard to the interaction of heat transfer and fluid flow in the melt pool during the production of Ti-6Al-4V components. The simulation results agreed reasonably well with the experimental data, in that 3 types of metal-transfer mode (liquid-bridge transition, droplet transition, intermediate) were predicted by simulations and confirmed by experiment. As the heat input was increased, the transfer mode changes from droplet transfer to liquid bridge. The latter was best, for ensuring good product quality, because of its stable dynamic equilibrium behavior: the metal transfer being driven mainly by the recoil pressure, while surface tension tends to disrupt the bridge. Interaction between these factors leads to oscillation of the liquid-bridge morphology during forming, with the oscillation frequency being about 200Hz. Droplet transfer occurs when the dynamic equilibrium is broken. It is here that the Marangoni effect plays an important role in droplet formation. It was deduced that the energy required to maintain the liquid bridge should be moderate, and that the heat-input conditions required for bridge maintenance should be quantitatively predictable.

Selective laser melting and wire arc additive manufacturing are competing metal additive-manufacturing techniques for the final shaping of small complex components and the near-final shaping of large components. The possibility of instead combining them has been considered[316]. Horizontal or vertical selective laser melting samples were used as a substrate for subsequent wire arc additive manufacturing. The results showed that Ti-6Al-4V samples which had been formed in this hybrid fashion typically consisted of 3 zones: a selective laser melting zone, a wire arc additive zone and an interface zone. Fine and short β-phase columnar grains which consisted mainly of α′-martensite existed in the selective laser melting zone. These fine grains then coarsened in the 3mm-thick interface zone, due to the repeated-heating regime. Coarse β-phase columnar grains replete with α lamellae led to epitaxial growth in the wire arc additive zone. Good metallurgical bonding existed in the interface-zone of both horizontal and vertical selective laser melting hybrids with wire arc additive manufacturing specimens, and all of the fractures occurred in the wire arc additive manufactured zone during tensile testing. The mean yield and ultimate tensile strengths of the horizontal selective laser melting hybrids with wire arc additive manufacture were 850 and 905MPa, respectively. The equivalent values for the vertical selective laser melting hybrids were 890 and 995MPa. All of the elongations were greater 10%, thus satisfying engineering requirements. The overall properties of the hybrids were comparable to, or better than, those of wire arc additive manufactured samples.

Although microstructural control of additively manufactured components is generally feasible by means of processing-parameter modification, this resource is limited in the case of materials such as Ti-6Al-4V. Grain-refinement and homogenization of cast titanium alloys can be achieved by adding hypoeutectic boron concentrations, and so the effect of adding up to 1.0wt% of boron during electron-beam melting has been explored[317]. Analysis of electron-beam fabricated material showed that the addition of 0.25 to 1.0wt% of boron progressively refined the grain structure and improved the hardness and elastic modulus. There was a reduction in β-phase size, but the grain structure remained columnar because of the directionality of the heat transfer during electron-beam manufacture.

In the case of as-printed selective laser melted Ti-6Al-4V, just a small (<1%) fraction of porosity can impair the properties and a large (>5%) fraction can damage the monotonic mechanical response[318]. The powder-size distribution and morphology are expected to affect power absorption spatially. From the macroscopic viewpoint, columnar grain structures predominate but rigorous control of thermal gradients and solidification rates can encourage the development of fully columnar, fully equiaxed or mixed structures. The preferred texture which is observed can be attributed mainly to preferential growth in the direction of the maximum thermal gradient corresponding to the selective laser melting direction. Due to the high cooling-rate, the β-phase transforms mainly to a very fine α'-martensite during cooling. As-prepared selective laser melt samples are thus associated with high ultimate strengths, high yield stress and relatively low ductilities although some further treatments can improve the balance between strength and ductility. Both the tensile strength and the ductility can be markedly improved by the formation of a duplex α-β microstructure during annealing. The tensile properties approach, and can exceed, those resulting from conventional processing. The presence of metallurgical defects involving the heterogeneity and anisotropy of microstructures limit the processing control of fracture behaviour.

Components made from Ti-6Al-4V, as fabricated by cold metal transfer additive manufacturing, are increasingly important in aerospace applications. An examination[319] of the microstructure and mechanical properties of as-prepared and heat-treated Ti-6Al-4V showed that as-prepared samples had a microstructure which consisted of acicular α'-martensite plus minor α-β lamellae, leading to a reasonable hardness and tensile strength. All of the α'-martensite could be transformed into α-β phase by treatment at 900C for 4h with furnace-cooling or by treatment at 1200C for 2h with furnace-cooling. By using the first or second heat-treatment, a higher hardness or a superior ductility, respectively, could be imparted to the cold metal transfer additively manufactured material. On the other hand, the tensile strength of as heat-treated material was appreciably lower than that of as-prepared samples.

The porosity which forms during layer-by-layer fabrication is a critical factor which may impair product performance. In the case of bulk cold spray prepared Ti-6Al-4V samples, hot isostatic pressing can be used to reduce the number of interior defects, modify the microstructure and improve the mechanical properties[320]. X-ray computer tomography shows that fully-dense Ti-6Al-4V can be obtained by high-temperature diffusion and high-pressure compaction. Following hot isostatic pressing, severely deformed grains underwent obvious growth, with uniformly distributed β precipitates arranged around equiaxed α grains. Tensile testing showed that the strength of cold spray Ti-6Al-4V could be greatly improved by enhanced diffusion and a consequent metallurgical bonding. Following hot isostatic pressing, the cold spray samples had a highly densified morphology which could improve the mechanical properties. This is something which is generally observed in the case of titanium. Thus hot isostatic pressing (110MPa, 1173K) of cold-sprayed pure titanium deposits markedly changes the microstructure, and reduces[321] the total porosity from 4.3 to 2.2% via the elimination of small-scale (less than 5μm) porosity, although larger pores are not completely closed. The ultimate tensile strengths of as-sprayed and hot isostatically pressed samples were 110 and 480MPa, respectively. The increased ultimate tensile strength can here too be attributed to mass diffusion and to microstructural changes which occur during pressing. On the other hand, the ductility of the pressed samples was only about 8% and therefore much lower than that of bulk pure titanium. Uniaxial fatigue data were correlated[322] with the effects of build orientation, surface roughness and hot isostatic pressing upon the properties of electron-beam prepared Ti-6Al-4V. Anisotropic fatigue behavior was observed in as-prepared components, and no statistical difference was found to exist between the fatigue performance of hot isostatically pressed and as-prepared material, for a given as-prepared surface condition. A fatigue life which was comparable to that of traditionally manufactured lamellar Ti-6Al-4V alloy was observed when both post-preparation hot isostatic pressing and machining were applied to electron beam melt-processed samples (figure 23).

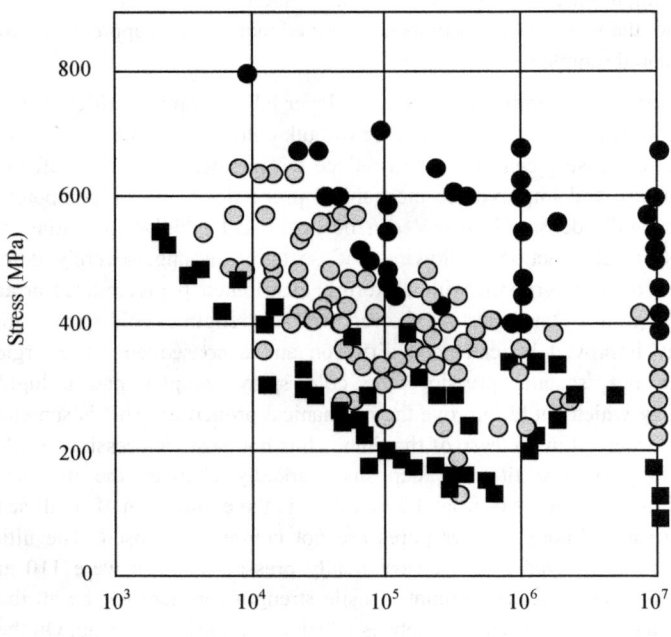

*Figure 23. Fatigue data for Ti-6Al-4V in various conditions:
conventionally processed (black circles), machined (grey circles), non-machined
(squares)*

The use of additional scans has been essayed[323] in an attempt to promote grain-coarsening by thus supplying additional thermal energy. It was supposed that the additional energy would cause a coarsening of the Ti-6Al-4V microstructure, thus improving the mechanical properties of an as-fabricated part as well as reducing surface porosity. Fatigue testing was therefore performed on an L-shaped bracket, using a loading configuration which was designed to cause failure at the corner. There was a 22% improvement in fatigue life, as compared with that of brackets in the as-fabricated condition or of brackets having a graded microstructure which had been created by selectively adding thermal energy to the predicted failure region. Three such brackets were exposed to a triple-melt cycle instead of the usual single-melt cycle, machined to a standard dimension and tested. The fatigue behaviour was close to that of a wrought L-shaped bracket.

An attempt has been made[324] to join a metal to a high-temperature polymer matrix composite by using additive manufacturing techniques. Thus, Ti-6Al-4V powder was connected by applying selective laser melting to carbon-fiber fabric, so that the latter's bottom layer was infused with polymer. The hybrid composite finally consisted of fully-densified metal on one face and a high-temperature polymer-matrix composite on the opposite face, together with a transition region which contained an appreciable amount of fiber-reinforcement connecting the faces.

Cold spray is a solid-state rapid-deposition method in which the metal powder is thrown, at supersonic velocities using a Laval nozzle, against a substrate. It can produce thick structures during additive manufacturing, without any melting. Image-analysis of particle-impact locations and ion-beam analysis of individual particles have supported[325] a 3-dimensional multi-component model of the process. The temperature and velocity variations for large particles, just before impact, were far less than those for small particles. An optimum particle-temperature and particle-velocity could therefore be found which ensured the greatest deformation upon impact and could thus improve the mechanical properties of cold-spray additively manufactured titanium parts. A computational fluid dynamics model was also developed[326] in order to compute the state of the supersonic gas which impinged on the substrate in the presence of surrounding air and to deduce the gas properties from the stagnation zone, through the nozzle and to the substrate surface, as well as the trajectory, velocity and temperature distribution of powder particles which were accelerated by the flow.

The existence of various sources of uncertainty in metal-based additive manufacturing processes prevents the production of additively manufactured parts of consistent quality. In the case of electron beam melting of Ti-6Al-4V, for example, it can be concluded that it would be necessary to establish a data-base of optimum processing conditions which relied upon physics-guided computer simulation. The main aim would be to identify those conditions which guarantee an equiaxed microstructure, in spite of the uncertainties. A 2-level data-driven model was proposed[327], based upon reliable design results, which indicated that a combination of low pre-heat temperature, low beam-power and intermediate scanning speeds maximises the probability of producing an equiaxed structure. Further analysis revealed that, among the 6 noise-factors which were considered, the specific heat capacity and the grain-growth activation energy had the greatest effects upon microstructure development.

Another uncertainty concerning product equality revolves around possible defects in additively manufactured parts. Given that acoustic methods have long been used for checking traditionally prepared samples, acoustic monitoring array was applied[328] to directed-energy deposited Ti-6Al-4V on a steel substrate. Waveforms were analyzed using temporal and spectral methods, and combined with metallographic studies. A comparison of crack densities with acoustic metrics confirmed the latter to be a satisfactory indicator of material damage.

The propagation of cracks in Ti-6Al-4V, produced using selective laser sintering, has in fact been studied[329] in some detail and compared with that in cold-rolled samples. The crack-propagation properties of additively manufactured specimens were similar to those of cold-rolled specimens.

Much of the original data on the limited number of alloys (Ti-6Al-4V, TiAl, Fe-Cr-Ni, Inconel-625, Inconel-718, Al-Si-10Mg) which have been the object of additive manufacturing have been summarised[330] with regard to hardness, tensile/compressive strength, fracture toughness and fatigue.

In a bonding experiment, tantalum was joined to titanium alloy[331]. Tantalum has a superior biocompatibility to that of other common orthopaedic alloys, and so tantalum-titanium hybrids could offer significant improvements. Bonding a tantalum coating to titanium is inherently difficult however because of the small difference between the melting point (3017C) of tantalum and the boiling point (3287C) of titanium. Laser powder deposition can melt a small volume of substrate, into which metal powder is sprayed, thus producing high temperatures and a high solidification rate. Laser deposition of Ti-Ta onto a Ti-6Al-4V substrate produced a solid surface and a structured coating having a pore size which was in the optimum range of 350 to 500µm.

In order to evaluate separately the creep and thermomechanical-fatigue behaviours of samples produced by selective laser melting, a novel specimen design was used[332] in which functional grading produced a coarse-grained core and a fine-grained outer shell. This arrangement allowed the best compromise to be made between the creep and fatigue performances.

As an example of an industrial application, a Ti-6Al-4V mini impeller was constructed[333] by means of laser additive manufacturing. The hardness, measured at various locations, revealed that the hub was stronger – with a maximum of 428HV – and exhibited less spatial variation as compared with the blades – with a maximum hardness of 415HV. There was also some anisotropy, with average hardness values of 397 and 385HV along the blade build direction and longitudinal direction, respectively. The hub bottom had a yield stress of 1193MPa, an ultimate tensile strength of 1310MPa and a total elongation of 5.5% in the longitudinal direction. The high strength was attributed to an entirely martensitic structure which had been produced by high cooling-rates at the build/substrate interface. The yield stress, ultimate tensile strength and total elongation along the build direction of a blade were 978MPa, 1096MPa and 9.12%, respectively, and the blade microstructure consisted of α' and α-β phases. As compared to polished blade specimens, unpolished specimens had a yield stress of 896 MPa, an ultimate tensile strength of 1018MPa and an elongation of 6.24%.

The electron-beam melting powder-bed additive process was used[334] to manufacture Ti-6Al-4V tensile specimens from pre-alloyed powders having particle sizes ranging from 45 to 100μm. The specimens had a zero degree orientation which was anticipated to exhibit a higher strength than that of other orientations. The room-temperature test results indicated a tensile strength of 1.2GPa and an elongation of about 8%. They showed that, for a zero degree manufacturing orientation, the yield stress and ultimate tensile strength were indeed higher than those usually found for wrought material.

Specimens made from Ti-6Al-4V were designed[335] in such a way as to permit accurate evaluation of factors such as energy-input, orientation and location. For a wide range of energy-inputs, with a speed factor of 30 to 40, small differences were found: such as a 2% change in the ultimate tensile strength and a 3% change in the yield stress. Vertically built parts exhibited no difference in ultimate tensile strength or yield stress, as compared to those of horizontally built parts, but the elongation at fracture was 30% lower. The difference elongation was attributed to the differing orientation of the tensile axis for horizontal and vertical parts. The orientation and location had a less than 3% effect on the mechanical properties. Other specimens were designed[336] so as to investigate the effects of processing parameters: such as the distance from the build plate, and the specimen size. The microstructure, and properties such as the microhardness, did not vary with distance from the build plate. The component-size did however influence the ultimate tensile strength and the yield stress, by less than 2%, over the size-range considered.

Electron beam selective melting has been used[337] to make Ti-6Al-4V parts from pre-alloyed powder. A Widmanstätten structure at the top of the product was associated with a slightly lower ultimate tensile strength, but a higher microhardness, than that of the bottom basket-weave microstructure. The ultimate tensile strength in the vertical orientation was lower than that in the horizontal orientation, and this was attributed to porosity defects caused by a lack of fusion. Overall, the ultimate tensile strength ranged from 923 to 1173MPa.

A study was made[338] of the properties of Ti-6Al-4V parts which had been created using selective laser melting. The parts were heat treated at 850C for 2h, with and without argon, in a vacuum furnace. The ultimate tensile strength of as-prepared parts was 1175MPa, while the value for samples treated under vacuum without argon was 750MPa. The ultimate tensile strength of samples heat-treated under vacuum with argon was 980MPa. The hardness of heat-treated samples was lower than the average Vickers hardness of as-prepared samples.

Plasma arc additive manufacturing has various advantages, such as convenience and cost, when compared with high-energy beam processes. Thin walls of Ti-6Al-4V were deposited[339] by means of wire-feed continuous plasma arc processing; the heat input bring decreased slightly from layer to layer. The thin walls exhibited various morphological features, including the epitaxial growth of prior β grains, horizontal layer bands, martensite and a basket-weave microstructure. Their appearance depended upon the heat input, thermal cycling and cooling-rate. By gradually reducing the heat input of each layer, and using a continuous current, the average yield stress, ultimate tensile strength and elongation could attain 877MPa, 968MPa and 1.5%, respectively. These values exceeded those which resulted from forging. The properties were improved due to a reduction of the aspect ratio of prior β grains and to a dispersion of particles among α-phase lamellae.

Microhardness and corrosion resistance studies of laser additively manufactured Ti-6Al-4V cladding layers in sulfuric acid solution revealed[340] maximum average microhardness of 390HV, a minimum corrosion current-density of $1.2337\mu A/cm^2$ and a maximum charge-transfer resistance of $11500\Omega/cm^2$ for a scanning-speed of 10mm/s. With increasing scanning-speed, the average microhardness had first increased and then decreased, while the corrosion current density had first decreased and then increased while the charge transfer resistance had first increased and then decreased. The main phase constituting the cladding layers was α-Ti, while fine acicular α' martensite formed near to the β grain boundary, giving an orthogonal basket-weave pattern.

Samples of Ti-6Al-4V plate were produced[341] by electron beam melting, and then friction stir-processed. An 80% reduction in the surface roughness was thus achieved, and the internal porosity was also decreased. The heating and stirring promoted the formation of distinct metallurgical zones: a stirred zone, a transition zone and a heat-affected zone. The stirred zone had a fully recrystallized microstructure, while heavily deformed non-recrystallized grains and coarse grains were found in the transition zone and heat-affected zone, respectively. Within the treated zone the hardness increased to 450HV, as compared with the 380HV of the base material.

*Table 29. Comparison of the compressive properties of selective laser melted and conventionally processed pure tungsten*

| Laser Energy (J/mm) | CYS (MPa) | UCS (MPa) | Strain (%) | Density (g/cm$^3$) | HV |
|---|---|---|---|---|---|
| 0.500 | 868 | 978 | 5.97 | 97.82 | 445 |
| 0.625 | 864 | 984 | 6.58 | 98.29 | 448 |
| 0.667 | 882 | 1015 | 6.76 | 98.50 | 461 |
| 0.750 | 791 | 933 | 8.65 | 97.98 | 452 |
| 0.833 | 849 | 964 | 6.64 | 97.93 | 467 |
| 1.000 | 860 | 962 | 6.36 | 97.72 | 456 |
| CVD | - | 780–1480 | ≤99.79 | - | 419 |
| HIP | 1010 | 1180 | ≤98.00 | - | - |
| PM | 900 | 1000–1200 | ≤98.20 | - | 344 |
| SPS | 750 | 980 | ≤96.30 | - | 372 |

Thin, 1.5 to 2.5mm, horizontal or vertical samples of selective laser melted Ti-6Al-4V plate and rolled plate have been used[342] as substrate materials for laser metal deposition. This led to the formation of a hybrid of the materials. The relative density of the hybrid could attain 99.5% due to the existence of pores having a diameter of up to 20μm. The tensile strength and elongation could reach 918MPa and 11%, respectively, and fracture occurred in the laser metal deposited zone. The occurrence of internal layer fracture of the latter zone increased elongation, while layer interface fracture decreased it. The laser deposition process epitaxially generated coarse columnar crystals, and laser re-melting reduced the microhardness of the selective laser melt substrate in a 2 to 3mm-thick heat-affected zone. The microhardness was 344, 343 and 375HV for the laser metal deposited zone, the heat-affected zone and substrate material, respectively, in the case of horizontally prepared selective laser melted material, 346, 334 and 386HV respectively in the case of vertically prepared selective laser melted material and 351, 328 and 340HV for rolled plate.

**Tungsten**

Due to the properties of tungsten, such as its high melting-point and high thermal conductivity, selective laser melting of pure tungsten parts encounters many challenges. Selective laser melting is nevertheless able to manufacture pure tungsten products having a density of $19.01g/cm^3$ (98.50% theoretical density) by optimizing the processing parameters[343]. A high-density microstructure was formed with no significant balling or macrocracks. Columnar grains stretched across several layers, and were parallel to the maximum temperature gradient, thus helping to ensure good bonding between the layers. The mechanical properties of these samples were comparable to those of tungsten produced using conventional methods, with a hardness exceeding $460HV_{0.05}$ and an ultimate compressive strength of about 1GPa (table 29). Highly spherical powders, sufficiently thin deposition layers a suitable linear laser energy were concluded to be essential when producing high-quality tungsten via selective laser melting. The laser linear energy had a marked effect upon the sintering formability, density and mechanical properties. The table shows that a linear energy of 0.667J/mm led to the best performance. Dense microstructures, without balling and with few microcracks, were formed. Even the few microcracks seemed not to affect seriously the performance. The effect of volumetric energy-density upon the densification behavior was such[344] that a maximum density of $19.0g/cm^3$, 98.4% of theoretical, was obtained by using an optimum energy-density of $1000J/mm^3$. The associated microstructure was moreover free from pores and balling. The microhardness and compressive strength of the selective laser-melted pure-tungsten product attained 474HV and 902MPa, respectively; comparable to the values found using conventional manufacturing methods. The morphology of the fracture surface indicated that the fracture mechanism was brittle fracture, with intergranular fracture being the main fracture path. Dry-sliding wear-tests indicated that the wear mechanism depended upon the energy-density which was used and, when the optimum processing conditions were used, adhesion of the hardened layers could occur. In that situation, the coefficient-of-friction was 0.45 and the wear-rate was $1.3 \times 10^{-5}mm^3/Nm$.

Solid-solution strengthened tungsten-matrix composites with nanocrystalline TiC have been prepared[345] by selective laser melting. A low laser linear energy-density led to the formation of micropores, while a high linear energy-density produced microcracks due to thermal stressing. Both eventualities tended to decrease selective laser melt densification. A uniform distribution of columnar crystals was produced, with an average diameter of $0.73\mu m$. At a moderate energy-density of 2.1kJ/m, processed composites had a 94.7% theoretical density and a microhardness of 8062MPa. Dry-sliding wear-tests indicated a coefficient of friction of 0.583, and a wear-rate of $6.9 \times 10^{-16}m^3/Nm$.

## Zinc

Laser powder-bed fusion has been used[346] to prepare Zn-2wt%Al alloy parts for bone-repair applications. A low energy-density led to the formation of pores, and thus an insufficient densification rate, due to the high liquid viscosity of the melt pool. A high energy-density caused evaporation of the zinc powder and thus failure of the process. With increasing energy-density, the resultant grains and the precipitated lamellar eutectic structure containing η-Zn and α-Al phases coarsened. This was attributed to increased heat accumulation and to the thereby diminished cooling rate. When using an optimum energy-density of 114.28J/mm$^3$, fully dense parts with a densification-rate of 98.3% were obtained, leading to a hardness of 64.5HV, a tensile strength of 192.2MPa and a corrosion rate of 0.14mm/y; with good biocompatibility.

A systematic study of selective laser melted zinc showed[347] that a single track surface following melting was rough and twisted, with a large number of attached particles, due to marked zinc evaporation. The relative density of the products was better than 99.50%. The surface roughness was 9.15 to 10.79μm in the as-melted condition, status and 4.83μm following sandblasting. The microstructure consisted of fine columnar grains and the average hardness, elastic modulus, yield stress, ultimate strength and elongation values were 42HV, 23GPa, 114MPa, 134MPa and 10.1%, respectively. These results were better than those usually obtained using conventional processes. This was attributed to the high densification and fine grain size.

Zinc-based alloys are suitable structural material for biodegradable applications. Laser powder-bed melting has been used[348] to produce porous pure zinc scaffolds with a macroscopic lattice-work appearance. Excessive evaporation of the high vapour-pressure metal during the laser-melting greatly affected the quality of the product. The scaffold surface was covered with many particles of various sizes. The average grain size across the width was 8.5μm and the hardness was 43.8HV. The zinc could also be combined[349] with the biodegradable magnesium alloy, WE43, in mass ratios of 2, 5 or 8%. A rapid cooling-rate, and the addition of WE43, resulted in grain refinement. With increasing WE43 content, more Zn-Mg$_2$Zn$_{11}$ eutectic precipitated, increasing the tensile strength but decreasing the elongation. Formation of the brittle MgZn$_2$ phase impaired the strength of Zn-8%WE43. The Zn-5%WE43 combination exhibited the highest (335.4MPa) tensile strength, but the elongation was only 1%. The compressive strength and Young's modulus of Zn-5%WE43 porous scaffolds were 73.2 and 2480MPa, respectively, while the equivalent values for the plain Zn scaffold were 22.9 and 950MPa. The overall properties were better than those obtained by most other manufacturing methods.

### Zirconium

Reactive *in situ* multi-material additive manufacturing was used[350] to integrate ceramic reinforcement into a zirconium-alloy via laser engineered net shaping. Good bonding was obtained between zirconium and Zr-BN composites on a Ti-6Al-4V substrate. Although the zirconium feedstock powder was α-phase, processing of the zirconium powder and Zr-BN powder led to the formation of some β-phase zirconium. The microstructure of the Zr-BN composites contained primary grains of zirconium diboride in a zirconium matrix. The hardness of the pure zirconium was 280HV. The hardness increased with an increasing BN content in the feedstock. In the case of the Zr-5%BN composite, the hardness was 421HV and that of the Zr-10%BN composite was 562HV. The Ti-6Al-4V substrate had a hardness of 223HV. The hardness of Zr-0%BN with no laser pass was 280HV. After a surface laser pass, the hardness increased to 326HV. For Zr-5%BN, after a laser pass, the hardness did not increase and was 428HV. The increase in hardness of samples with no BN additions, and exposed only due to a post-deposition laser scan was attributed to an increase in residual stresses arising from the laser heat-input, followed by rapid solidification. There was no decrease in the size of the grains. The increase in hardness upon adding 5%BN was attributed mainly to the formation of zirconium diboride ceramic in the metal matrix.

### References

[1] R.Baker, US Patent No. 1533300, 14th April 1925.

[2] Boyce, B.L., Salzbrenner, B.C., Rodelas, J.M., Swiler, L.P., Madison, J.D., Jared, B.H., Shen, Y.L., Advanced Engineering Materials, 19[8] 2017, 1700102. https://doi.org/10.1002/adem.201700102

[3] Ding, X., Du, F., Zhang, Q., Wang, G., Fan, S., Duan, X., Proceedings of SPIE, 10842, 2019, 108420L.

[4] Lotz, C., Wessarges, Y., Hermsdorf, J., Ertmer, W., Overmeyer, L., Advances in Space Research, 61[8] 2018, 1967-1974. https://doi.org/10.1016/j.asr.2018.01.010

[5] Gu, H., Li, L., International Journal of Heat and Mass Transfer, 140, 2019, 51-65. https://doi.org/10.1016/j.ijheatmasstransfer.2019.05.081

[6] Goulas, A., Binner, J.G.P., Engstrøm, D.S., Harris, R.A., Friel, R.J., Proceedings of the Institution of Mechanical Engineers - L, 233[8] 2019, 1629-1644. https://doi.org/10.1177/1464420718777932

[7] Yuqing, M., Liming, K., Chunping, H., Fencheng, L., Qiang, L., International Journal of Advanced Manufacturing Technology, 83[9-12] 2016, 1637-1647.

https://doi.org/10.1007/s00170-015-7695-9

[8] Deshpande, A., Hsu, K., Additive Manufacturing, 19, 2018, 73-80. https://doi.org/10.1016/j.addma.2017.11.006

[9] Wang, B., Zhang, H., Zhu, X., Wang, Y., Jin, W., He, P., Journal of Mechanical Engineering, 54[22] 2018, 95-102. https://doi.org/10.3901/JME.2018.22.095

[10] Tariq, N.H., Gyansah, L., Qiu, X., Du, H., Wang, J.Q., Feng, B., Yan, D.S., Xiong, T.Y., Materials and Design, 156, 2018, 287-299. https://doi.org/10.1016/j.matdes.2018.06.062

[11] Tariq, N.U.H., Gyansah, L., Qiu, X., Jia, C., Awais, H.B., Zheng, C., Du, H., Wang, J., Xiong, T., Journal of Materials Science and Technology, 35[6] 2019, 1053-1063. https://doi.org/10.1016/j.jmst.2018.12.022

[12] Gyansah, L., Tariq, N.H., Tang, J.R., Qiu, X., Gao, J.Z., Wang, J.Q., Xiong, T.Y., Materials Science Forum, 932, 2018, 62-75. https://doi.org/10.4028/www.scientific.net/MSF.932.62

[13] Su, J., Hubbard, W., Dubrowsky, J., Griffiths, R.J., Yu, H.Z., Hardwick, N., Proceedings of the International Conference on Powder Metallurgy and Particulate Materials, 2017, 687-692.

[14] Griffiths, R.J., Perry, M.E.J., Sietins, J.M., Zhu, Y., Hardwick, N., Cox, C.D., Rauch, H.A., Yu, H.Z., Journal of Materials Engineering and Performance, 28[2] 2019, 648-656. https://doi.org/10.1007/s11665-018-3649-3

[15] Manca, D.R., Churyumov, A.Y., Pozdniakov, A.V., Prosviryakov, A.S., Ryabov, D.K., Krokhin, A.Y., Korolev, V.A., Daubarayte, D.K., Metals and Materials International, 25[3] 2019, 633-640. https://doi.org/10.1007/s12540-018-00211-0

[16] Ma, T., Ge, J., Chen, Y., Jin, T., Lei, Y., Materials Letters, 237, 2019, 266-269. https://doi.org/10.1016/j.matlet.2018.11.115

[17] Cong, B., Qi, Z., Qi, B., Sun, H., Zhao, G., Ding, J., Applied Sciences, 7[3] 2017, 275. https://doi.org/10.3390/app7030275

[18] Qi, Z., Cong, B., Qi, B., Sun, H., Zhao, G., Ding, J., Journal of Materials Processing Technology, 255, 2018, 347-353. https://doi.org/10.1016/j.jmatprotec.2017.12.019

[19] Polizzi, A.J., Iten, J.J., Materials Science and Technology 2018, 26-32.

[20] Sun, H., Cong, B., Qi, Z., Qi, B., Zhao, G., Ding, J., Rare Metal Materials and Engineering, 46[8] 2017, 2203-2207.

[21] Gu, J., Bai, J., Ding, J., Williams, S., Wang, L., Liu, K., Journal of Materials

Processing Technology, 262, 2018, 210-220.
https://doi.org/10.1016/j.jmatprotec.2018.06.030

[22] Wang, L., Suo, Y., Liang, Z., Wang, D., Wang, Q., Materials Letters, 241, 2019, 231-234. https://doi.org/10.1016/j.matlet.2019.01.117

[23] Chou, R., Milligan, J., Paliwal, M., Brochu, M., JOM, 67[3] 2015, 590-596. https://doi.org/10.1007/s11837-014-1272-9

[24] Feng, Y., He, L., Wang, K., Xuanyu, E., Journal of Materials Engineering and Performance, 27[11] 2018, 5591-5604. https://doi.org/10.1007/s11665-018-3710-2

[25] Zhang, C., Gao, M., Zeng, X., Journal of Materials Processing Technology, 271, 2019, 85-92. https://doi.org/10.1016/j.jmatprotec.2019.03.028

[26] Gu, J., Wang, X., Bai, J., Ding, J., Williams, S., Zhai, Y., Liu, K., Materials Science and Engineering A, 712, 2018, 292-301.
https://doi.org/10.1016/j.msea.2017.11.113

[27] Horgar, A., Fostervoll, H., Nyhus, B., Ren, X., Eriksson, M., Akselsen, O.M., Journal of Materials Processing Technology, 259, 2018, 68-74.
https://doi.org/10.1016/j.jmatprotec.2018.04.014

[28] Fang, X., Zhang, L., Chen, G., Dang, X., Huang, K., Wang, L., Lu, B., Materials, 11[11] 2018, 2075. https://doi.org/10.3390/ma11112075

[29] Zhang, B., Zhang, L., Wang, C., Wang, Z., Gao, Q., Journal of Materials Processing Technology, 267, 2019, 167-176. https://doi.org/10.1016/j.jmatprotec.2018.12.011

[30] Geng, H., Li, J., Xiong, J., Lin, X., Zhang, F., Journal of Materials Engineering and Performance, 26[2] 2017, 621-629. https://doi.org/10.1007/s11665-016-2480-y

[31] Huang, D., Zhu, Z.H., Geng, H.B., Xiong, J.T., Li, J.L., Zhang, F.S., Journal of Materials Engineering, 45[3] 2017, 66-72.

[32] Li, R., Chen, H., Zhu, H., Wang, M., Chen, C., Yuan, T., Materials and Design, 168, 2019, 107668. https://doi.org/10.1016/j.matdes.2019.107668

[33] Zhang, H., Gu, D., Yang, J., Dai, D., Zhao, T., Hong, C., Gasser, A., Poprawe, R., Additive Manufacturing, 23, 2018, 1-12.
https://doi.org/10.1016/j.addma.2018.07.002

[34] Uddin, S.Z., Murr, L.E., Terrazas, C.A., Morton, P., Roberson, D.A., Wicker, R.B., Additive Manufacturing, 22, 2018, 405-415.
https://doi.org/10.1016/j.addma.2018.05.047

[35] Sridharan, N., Wolcott, P., Dapino, M., Babu, S.S., Science and Technology of

Welding and Joining, 22[5] 2017, 373-380.
https://doi.org/10.1080/13621718.2016.1249644

[36] Hahnlen, R., Dapino, M.J., ASME Conference on Smart Materials, Adaptive Structures and Intelligent Systems, 2, 2012, 505-514.

[37] Wolcott, P.J., Hehr, A., Dapino, M.J., Proceedings of SPIE, 9059, 2014, 905908.

[38] Prashanth, K.G., Scudino, S., Eckert, J., Acta Materialia, 126, 2017, 25-35.
https://doi.org/10.1016/j.actamat.2016.12.044

[39] Yang, Q., Xia, C., Deng, Y., Materials Science Forum, 944, 2018, 64-72.
https://doi.org/10.4028/www.scientific.net/MSF.944.64

[40] Luo, X., Han, Y., Li, Q., Hu, X., Xue, L., Journal of Wuhan University of Technology, 33[3] 2018, 715-719. https://doi.org/10.1007/s11595-018-1883-z

[41] Heard, D.W., Brophy, S., Brochu, M., Canadian Metallurgical Quarterly, 51[3] 2012, 302-312. https://doi.org/10.1179/1879139512Y.0000000023

[42] Kurz, W., Fisher, D.J., Fundamentals of Solidification, Trans Tech Publications, 1998. https://doi.org/10.4028/www.scientific.net/RC.35

[43] Demir, A.G., Biffi, C.A., Journal of Manufacturing Processes, 37, 2019, 362-369.
https://doi.org/10.1016/j.jmapro.2018.11.017

[44] Zhang, M., Chen, C., Huang, Y., Materials Science and Technology, 34[8] 2018, 968-981. https://doi.org/10.1080/02670836.2017.1415014

[45] Qiu, X., Tariq, N.U.H., Qi, L., Zan, Y.N., Wang, Y.J., Wang, J.Q., Du, H., Xiong, T.Y., Journal of Alloys and Compounds, 780, 2019, 597-606.
https://doi.org/10.1016/j.jallcom.2018.11.399

[46] Kimura, T., Nakamoto, T., Journal of the Japan Society of Powder and Powder Metallurgy, 61[11] 2014, 531-537. https://doi.org/10.2497/jjspm.61.531

[47] Muñiz-Lerma, J.A., Nommeots-Nomm, A., Waters, K.E., Brochu, M., Materials, 11[12] 2018, 2386. https://doi.org/10.3390/ma11122386

[48] Buchbinder, D., Meiners, W., Pirch, N., Wissenbach, K., Schrage, J., Journal of Laser Applications, 26[1] 2014, 012004. https://doi.org/10.2351/1.4828755

[49] Uzan, N.E., Shneck, R., Yeheskel, O., Frage, N., Additive Manufacturing, 24, 2018, 257-263. https://doi.org/10.1016/j.addma.2018.09.033

[50] Hitzler, L., Janousch, C., Schanz, J., Merkel, M., Heine, B., Mack, F., Hall, W., Öchsner, A., Journal of Materials Processing Technology, 243, 2017, 48-61.

https://doi.org/10.1016/j.jmatprotec.2016.11.029

[51] Liu, J., To, A.C., Additive Manufacturing, 16, 2017, 58-64.
https://doi.org/10.1016/j.addma.2017.05.005

[52] Chen, B., Moon, S.K., Yao, X., Bi, G., Shen, J., Umeda, J., Kondoh, K., JOM, 70[5]
2018, 644-649. https://doi.org/10.1007/s11837-018-2793-4

[53] Chen, B., Yao, Y., Song, X., Tan, C., Cao, L., Feng, J., Ferroelectrics, 523[1] 2018,
153-166. https://doi.org/10.1080/00150193.2018.1392147

[54] Ferro, C.G., Varetti, S., Maggiore, P., Lombardi, M., Biamino, S., Manfredi, D.,
Calignano, F., Journal of Sandwich Structures and Materials, 2018, in press.

[55] Del Re, F., Contaldi, V., Astarita, A., Palumbo, B., Squillace, A., Corrado, P., Di
Petta, P., International Journal of Advanced Manufacturing Technology, 97[5-8]
2018, 2231-2240. https://doi.org/10.1007/s00170-018-2090-y

[56] Girelli, L., Tocci, M., Montesano, L., Gelfi, M., Pola, A., IOP Conference Series -
Materials Science and Engineering, 264[1] 2017, 012016.
https://doi.org/10.1088/1757-899X/264/1/012016

[57] Gu, D., Chang, F., Dai, D., Journal of Manufacturing Science and Engineering,
Transactions of the ASME, 137[2] 2015, 021010.
https://doi.org/10.1115/1.4028925

[58] Aboulkhair, N.T., Maskery, I., Tuck, C., Ashcroft, I., Everitt, N.M., Materials
Science and Engineering A, 667, 2016, 139-146.
https://doi.org/10.1016/j.msea.2016.04.092

[59] Javidani, M., Arreguin-Zavala, J., Danovitch, J., Tian, Y., Brochu, M., Journal of
Thermal Spray Technology, 26[4] 2017, 587-597. https://doi.org/10.1007/s11666-
016-0495-4

[60] Li, C., Sun, S., Zhang, Y., Liu, C., Deng, P., Zeng, M., Wang, F., Ma, P., Li, W.,
Wang, Y., International Journal of Advanced Manufacturing Technology, 103[5-8]
2019, 3235-3246. https://doi.org/10.1007/s00170-019-04001-9

[61] Yu, T., Hyer, H., Sohn, Y., Bai, Y., Wu, D., Materials and Design, 182, 2019,
108062. https://doi.org/10.1016/j.matdes.2019.108062

[62] Singh, A., Ramakrishnan, A., Baker, D., Biswas, A., Dinda, G.P., Journal of Alloys
and Compounds, 719, 2017, 151-158.
https://doi.org/10.1016/j.jallcom.2017.05.171

[63] Casati, R., Coduri, M., Riccio, M., Rizzi, A., Vedani, M., Journal of Alloys and

Compounds, 801, 2019, 243-253. https://doi.org/10.1016/j.jallcom.2019.06.123

[64] Zuo, H., Li, H., Qi, L., Zhong, S., Journal of Materials Science and Technology, 32[5] 2016, 485-488. https://doi.org/10.1016/j.jmst.2016.03.004

[65] Sistla, H., Newkirk, J.W., Liou, F.F., Materials Science Forum, 783-786, 2014, 2370-2375. https://doi.org/10.4028/www.scientific.net/MSF.783-786.2370

[66] Sistla, H.R., Newkirk, J.W., Liou, F.F., Materials and Design, 81, 2015, 113-121. https://doi.org/10.1016/j.matdes.2015.05.027

[67] Gwalani, B., Soni, V., Waseem, O.A., Mantri, S.A., Banerjee, R., Optics and Laser Technology, 113, 2019, 330-337. https://doi.org/10.1016/j.optlastec.2019.01.009

[68] Schubert, T., Breninek, A., Bernthaler, T., Sellmer, D., Schneider, M., Schneider, G., Practical Metallography, 54[9] 2017, 577-595. https://doi.org/10.3139/147.110477

[69] Zhang, X., Guo, Z., Chen, C., Yang, W., International Journal of Refractory Metals and Hard Materials, 70, 2018, 215-223. https://doi.org/10.1016/j.ijrmhm.2017.10.005

[70] Davydova, A., Domashenkov, A., Sova, A., Movtchan, I., Bertrand, P., Desplanques, B., Peillon, N., Saunier, S., Desrayaud, C., Bucher, S., Iacob, C., Journal of Materials Processing Technology, 229, 2016, 361-366. https://doi.org/10.1016/j.jmatprotec.2015.09.033

[71] Hitzler, L., Alifui-Segbaya, F., Williams, P., Heine, B., Heitzmann, M., Hall, W., Merkel, M., Öchsner, A., Advances in Materials Science and Engineering, 2018, 8213023. https://doi.org/10.1155/2018/8213023

[72] Ghani, S.A.C., Mohamed, S.R., Harun, W.S.W., Noar, N.A.Z.M., AIP Conference Proceedings, 1901, 2017, 100001.

[73] Kenel, C., Casati, N.P.M., Dunand, D.C., Nature Communications, 10[1] 2019, 904. https://doi.org/10.1038/s41467-019-08763-4

[74] Gaytan, S.M., Murr, L.E., Martinez, E., Martinez, J.L., MacHado, B.I., Ramirez, D.A., Medina, F., Collins, S., Wicker, R.B., Metallurgical and Materials Transactions A: Physical Metallurgy and Materials Science, 41[12] 2010, 3216-3227. https://doi.org/10.1007/s11661-010-0388-y

[75] Mantrala, K.M., Das, M., Balla, V.K., Srinivasa Rao, C., Kesava Rao, V.V.S.,) Journal of Materials Research, 29[17] 2014, 2021-2027. https://doi.org/10.1557/jmr.2014.163

[76] Sang, J., Wang, B., Zhu, X., Zhang, H., Wang, Y., Jin, W., Gao, B., Chang, Q., He,

P., Materials Review, 32[9] 2018, 3199-3207.

[77] Kumar, A.Y., Wang, J., Bai, Y., Huxtable, S.T., Williams, C.B., Materials and Design, 182, 2019, 108001. https://doi.org/10.1016/j.matdes.2019.108001

[78] Hu, Z., Chen, F., Lin, D., Nian, Q., Parandoush, P., Zhu, X., Shao, Z., Cheng, G.J., Nanotechnology, 28[44] 2017, 445705. https://doi.org/10.1088/1361-6528/aa8946

[79] Ramirez, D.A., Murr, L.E., Martinez, E., Hernandez, D.H., Martinez, J.L., MacHado, B.I., Medina, F., Frigola, P., Wicker, R.B., Acta Materialia, 59[10] 2011, 4088-4099. https://doi.org/10.1016/j.actamat.2011.03.033

[80] Dong, B., Pan, Z., Shen, C., Ma, Y., Li, H., Metallurgical and Materials Transactions B, 48[6] 2017, 3143-3151. https://doi.org/10.1007/s11663-017-1071-0

[81] Mao, Z., Zhang, D.Z., Wei, P., Zhang, K., Materials, 10[4] 2017, 333. https://doi.org/10.3390/ma10040333

[82] Popovich, A., Sufiiarov, V., Polozov, I., Borisov, E., Masaylo, D., Orlov, A., Materials Letters, 179, 2016, 38-41. https://doi.org/10.1016/j.matlet.2016.05.064

[83] Sabelle, M., Walczak, M., Ramos-Grez, J., Optics and Lasers in Engineering, 100, 2018, 1-8. https://doi.org/10.1016/j.optlaseng.2017.06.028

[84] Mao, Z., Zhang, D.Z., Jiang, J., Fu, G., Zhang, P., Materials Science and Engineering A, 721, 2018, 125-134. https://doi.org/10.1016/j.msea.2018.02.051

[85] Wilms, M.B., Streubel, R., Frömel, F., Weisheit, A., Tenkamp, J., Walther, F., Barcikowski, S., Schleifenbaum, J.H., Gökce, B., Procedia CIRP, 74, 2018, 196-200. https://doi.org/10.1016/j.procir.2018.08.093

[86] Aucott, L., Dong, H., Mirihanage, W., Atwood, R., Kidess, A., Gao, S., Wen, S., Marsden, J., Feng, S., Tong, M., Connolley, T., Drakopoulos, M., Kleijn, C.R., Richardson, I.M., Browne, D.J., Mathiesen, R.H., Atkinson, H.V., Nature Communications, 9[1] 2018, 5414. https://doi.org/10.1038/s41467-018-07900-9

[87] Hu, X., Nycz, A., Lee, Y., Shassere, B., Simunovic, S., Noakes, M., Ren, Y., Sun, X., Materials Science and Engineering A, 761, 2019, 138057. https://doi.org/10.1016/j.msea.2019.138057

[88] Youheng, F., Guilan, W., Haiou, Z., Liye, L., International Journal of Advanced Manufacturing Technology, 91[1-4] 2017, 301-313. https://doi.org/10.1007/s00170-016-9621-1

[89] Zeng, G.H., Song, T., Dai, Y.H., Tang, H.P., Yan, M., Materials and Design, 169, 2019, 107693. https://doi.org/10.1016/j.matdes.2019.107693

[90] Levy, A., Miriyev, A., Elliott, A., Babu, S.S., Frage, N., Materials and Design, 118, 2017, 198-203. https://doi.org/10.1016/j.matdes.2017.01.024

[91] Liu, L., Zhuang, Z., Liu, F., Zhu, M., International Journal of Advanced Manufacturing Technology, 69[9-12] 2013, 2131-2137. https://doi.org/10.1007/s00170-013-5191-7

[92] Wang, Y., Shi, J., Lu, S., Xiao, W., ASME 12th International Manufacturing Science and Engineering Conference, 2017, 2.

[93] Niendorf, T., Brenne, F., Hoyer, P., Schwarze, D., Schaper, M., Grothe, R., Wiesener, M., Grundmeier, G., Maier, H.J., Metallurgical and Materials Transactions A, 46[7] 2015, 2829-2833. https://doi.org/10.1007/s11661-015-2932-2

[94] Shen, C., Pan, Z., Cuiuri, D., Dong, B., Li, H., Materials Science and Engineering A, 669, 2016, 118-126. https://doi.org/10.1016/j.msea.2016.05.047

[95] Shen, C., Pan, Z., Cuiuri, D., Roberts, J., Li, H., Metallurgical and Materials Transactions B, 47[1] 2016, 763-772. https://doi.org/10.1007/s11663-015-0509-5

[96] Ge, J., Lin, J., Lei, Y., Fu, H., Materials Science and Engineering A, 715, 2018, 144-153. https://doi.org/10.1016/j.msea.2017.12.076

[97] Ge, J., Lin, J., Chen, Y., Lei, Y., Fu, H., Journal of Alloys and Compounds, 748, 2018, 911-921. https://doi.org/10.1016/j.jallcom.2018.03.222

[98] Ge, J., Lin, J., Fu, H., Lei, Y., Xiao, R., Materials Letters, 232, 2018, 11-13. https://doi.org/10.1016/j.matlet.2018.08.037

[99] Ge, J., Lin, J., Fu, H., Lei, Y., Xiao, R., Materials and Design, 160, 2018, 218-228. https://doi.org/10.1016/j.matdes.2018.09.021

[100] Sundqvist, J., Kaplan, A.F.H., Optics and Laser Technology, 108, 2018, 487-495. https://doi.org/10.1016/j.optlastec.2018.07.031

[101] Abd-Elghany, K., Bourell, D.L., Rapid Prototyping Journal, 18[5] 2012, 420-428. https://doi.org/10.1108/13552541211250418

[102] Huang, W., Zhang, Y., Dai, W., Long, R., Materials Science and Engineering A, 758, 2019, 60-70. https://doi.org/10.1016/j.msea.2019.04.108

[103] Haden, C.V., Zeng, G., Carter, F.M., III, Ruhl, C., Krick, B.A., Harlow, D.G., Additive Manufacturing, 16, 2017, 115-123. https://doi.org/10.1016/j.addma.2017.05.010

[104] Prado-Cerqueira, J.L., Camacho, A.M., Diéguez, J.L., Rodríguez-Prieto, Á.,

Aragón, A.M., Lorenzo-Martín, C., Yanguas-Gil, Á., Materials, 11[8] 2018, 1449. https://doi.org/10.3390/ma11081449

[105] Gualtieri, T., Bandyopadhyay, A., Materials and Design, 139, 2018, 419-428. https://doi.org/10.1016/j.matdes.2017.11.007

[106] Gordon, J.V., Haden, C.V., Nied, H.F., Vinci, R.P., Harlow, D.G., Materials Science and Engineering A, 724, 2018, 431-438. https://doi.org/10.1016/j.msea.2018.03.075

[107] Cavaleiro, A.J., Fernandes, C.M., Farinha, A.R., Gestel, C.V., Jhabvala, J., Boillat, E., Senos, A.M.R., Vieira, M.T., Applied Surface Science, 427, 2018, 131-138. https://doi.org/10.1016/j.apsusc.2017.08.039

[108] Bai, Y., Gao, Q., Chen, X., Yin, H., Fang, L., Zhao, J., IOP Conference Series - Earth and Environmental Science, 237[3] 2019, 032096. https://doi.org/10.1088/1755-1315/237/3/032096

[109] Cherry, J.A., Davies, H.M., Mehmood, S., Lavery, N.P., Brown, S.G.R., Sienz, J., International Journal of Advanced Manufacturing Technology, 76[5-8] 2014, 869-879. https://doi.org/10.1007/s00170-014-6297-2

[110] Xu, X., Mi, G., Luo, Y., Jiang, P., Shao, X., Wang, C., Optics and Lasers in Engineering, 94, 2017, 1-11. https://doi.org/10.1016/j.optlaseng.2017.02.008

[111] Heer, B., Bandyopadhyay, A., Materials Letters, 216, 2018, 16-19. https://doi.org/10.1016/j.matlet.2017.12.129

[112] Gordon, J.V., Vinci, R.P., Hochhalter, J.D., Rollett, A.D., Gary Harlow, D., Materialia, 7, 2019, 100397. https://doi.org/10.1016/j.mtla.2019.100397

[113] Akbari, M., Kovacevic, R., Additive Manufacturing, 23, 2018, 487-497. https://doi.org/10.1016/j.addma.2018.08.031

[114] Li, Z., Voisin, T., McKeown, J.T., Ye, J., Braun, T., Kamath, C., King, W.E., Wang, Y.M., International Journal of Plasticity, 120, 2019, 395-410. https://doi.org/10.1016/j.ijplas.2019.05.009

[115] Chen, J., Yang, Y., Song, C., Zhang, M., Wu, S., Wang, D., Materials Science and Engineering A, 752, 2019, 75-85. https://doi.org/10.1016/j.msea.2019.02.097

[116] Wu, A.S., Brown, D.W., Kumar, M., Gallegos, G.F., King, W.E., Metallurgical and Materials Transactions A, 45[13] 2014, 6260-6270. https://doi.org/10.1007/s11661-014-2549-x

[117] Kamariah, M.S.I.N., Harun, W.S.W., Khalil, N.Z., Ahmad, F., Ismail, M.H., Sharif,

S., IOP Conference Series - Materials Science and Engineering, 257[1] 2017, 012021. https://doi.org/10.1088/1757-899X/257/1/012021

[118] Miao, P., Niu, F., Ma, G., Lü, J., Wu, D., Opto-Electronic Engineering, 44[4] 2017, 410-417.

[119] Zheng, Z., Wang, L., Jia, M., Cheng, L., Yan, B., Materials Science and Engineering C, 71, 2017, 1099-1105. https://doi.org/10.1016/j.msec.2016.11.032

[120] Sun, Z., Tan, X., Tor, S.B., Yeong, W.Y., Materials and Design, 104, 2016, 197-204. https://doi.org/10.1016/j.matdes.2016.05.035

[121] Laohaprapanon, A., Jeamwatthanachai, P., Wongcumchang, M., Chantarapanich, N., Chantaweroad, S., Sitthiseripratip, K., Wisutmethangoon, S., Advanced Materials Research, 341-342, 2012, 816-820. https://doi.org/10.4028/www.scientific.net/AMR.341-342.816

[122] Kalita, S., Materials Science and Technology, 2011, 709-717.

[123] Kalita, S., Materials Science and Technology Conference and Exhibition, 2, 2011, 1485-1492.

[124] Sasaki, T., Iwatsuki, H., Yamaguchi, T., Yamaguchi, D., International Conference on Digital Printing Technologies, 2016, 139-142.

[125] Zhong, Y., Rännar, L.E., Liu, L., Koptyug, A., Wikman, S., Olsen, J., Cui, D., Shen, Z., Journal of Nuclear Materials, 486, 2017, 234-245. https://doi.org/10.1016/j.jnucmat.2016.12.042

[126] Yang, N., Yee, J., Zheng, B., Gaiser, K., Reynolds, T., Clemon, L., Lu, W.Y., Schoenung, J.M., Lavernia, E.J., Journal of Thermal Spray Technology, 26[4] 2017, 610-626. https://doi.org/10.1007/s11666-016-0480-y

[127] Roehling, T.T., Wu, S.S.Q., Khairallah, S.A., Roehling, J.D., Soezeri, S.S., Crumb, M.F., Matthews, M.J., Acta Materialia, 128, 2017, 197-206. https://doi.org/10.1016/j.actamat.2017.02.025

[128] Chen, X., Li, J., Cheng, X., He, B., Wang, H., Huang, Z., Materials Science and Engineering A, 703, 2017, 567-577. https://doi.org/10.1016/j.msea.2017.05.024

[129] Chen, X., Li, J., Cheng, X., Wang, H., Huang, Z., Materials Science and Engineering A, 715, 2018, 307-314. https://doi.org/10.1016/j.msea.2017.10.002

[130] AlMangour, B., Kim, Y.K., Grzesiak, D., Lee, K.A., Composites B, 156, 2019, 51-63. https://doi.org/10.1016/j.compositesb.2018.07.050

[131] Haley, J.C., Schoenung, J.M., Lavernia, E.J., Materials Science and Engineering A,

761, 2019, 138052. https://doi.org/10.1016/j.msea.2019.138052

[132] Gong, H., Snelling, D., Kardel, K., Carrano, A., JOM, 71[3] 2019, 880-885. https://doi.org/10.1007/s11837-018-3207-3

[133] Korinko, P.S., Bobbitt, J.T., Morgan, M.J., Reigel, M., List, F.A., Babu, S.S., JOM, 71[3] 2019, 1095-1104. https://doi.org/10.1007/s11837-019-03341-x

[134] Marya, M., Singh, V., Marya, S., Hascoet, J.Y., Metallurgical and Materials Transactions B, 46[4] 2015, 1654-1665. https://doi.org/10.1007/s11663-015-0310-5

[135] Segura, I.A., Murr, L.E., Terrazas, C.A., Bermudez, D., Mireles, J., Injeti, V.S.V., Li, K., Yu, B., Misra, R.D.K., Wicker, R.B., Journal of Materials Science and Technology, 35[2] 2019, 351-367. https://doi.org/10.1016/j.jmst.2018.09.059

[136] Chen, W., Yin, G., Feng, Z., Liao, X., Metals, 8[9] 2018, 729. https://doi.org/10.3390/met8090729

[137] Yin, S., Chen, C., Yan, X., Feng, X., Jenkins, R., O'Reilly, P., Liu, M., Li, H., Lupoi, R., Additive Manufacturing, 22, 2018, 592-600. https://doi.org/10.1016/j.addma.2018.06.005

[138] Yan, X., Huang, C., Chen, C., Bolot, R., Dembinski, L., Huang, R., Ma, W., Liao, H., Liu, M., Surface and Coatings Technology, 371, 2019, 161-171. https://doi.org/10.1016/j.surfcoat.2018.03.072

[139] Tan, C., Zhou, K., Kuang, M., Ma, W., Kuang, T., Science and Technology of Advanced Materials, 19[1] 2018, 746-758. https://doi.org/10.1080/14686996.2018.1527645

[140] Meneghetti, G., Rigon, D., Cozzi, D., Waldhauser, W., Dabalà, M., Procedia Structural Integrity, 7, 2017, 149-157. https://doi.org/10.1016/j.prostr.2017.11.072

[141] Mutua, J., Nakata, S., Onda, T., Chen, Z.C., Materials and Design, 139, 2018, 486-497. https://doi.org/10.1016/j.matdes.2017.11.042

[142] Kürnsteiner, P., Wilms, M.B., Weisheit, A., Barriobero-Vila, P., Jägle, E.A., Raabe, D., Acta Materialia, 129, 2017, 52-60. https://doi.org/10.1016/j.actamat.2017.02.069

[143] Casati, R., Coduri, M., Lecis, N., Andrianopoli, C., Vedani, M., Materials Characterization, 137, 2018, 50-57. https://doi.org/10.1016/j.matchar.2018.01.015

[144] Hentschel, O., Siegel, L., Scheitler, C., Huber, F., Junker, D., Gorunow, A., Schmidt, M., Metals, 8[9] 2018, 659. https://doi.org/10.3390/met8090659

[145] Ali, Y., Henckell, P., Hildebrand, J., Reimann, J., Bergmann, J.P., Barnikol-Oettler, S., Journal of Materials Processing Technology, 269, 2019, 109-116. https://doi.org/10.1016/j.jmatprotec.2019.01.034

[146] Chen, C.J., Yan, K., Qin, L., Zhang, M., Wang, X., Zou, T., Hu, Z., Journal of Materials Engineering and Performance, 26[11] 2017, 5577-5589. https://doi.org/10.1007/s11665-017-2992-0

[147] Ou, W., Mukherjee, T., Knapp, G.L., Wei, Y., DebRoy, T., International Journal of Heat and Mass Transfer, 127, 2018, 1084-1094. https://doi.org/10.1016/j.ijheatmasstransfer.2018.08.111

[148] Jung, I.D., Choe, J., Yun, J., Yang, S., Yang, D.Y., Kim, Y.J., Yu, J.H., Archives of Metallurgy and Materials, 64[2] 2019, 571-578.

[149] Zhang, X., Li, W., Chen, X., Cui, W., Liou, F., International Journal of Advanced Manufacturing Technology, 95[9-12] 2018, 3335-3348. https://doi.org/10.1007/s00170-017-1455-y

[150] An, W., Park, J., Lee, J., Choe, J., Jung, I.D., Yu, J.H., Kim, S., Sung, H., Korean Journal of Materials Research, 28[11] 2018, 663-670. https://doi.org/10.3740/MRSK.2018.28.11.663

[151] Körperich, J.P., Merkel, M., Materialwissenschaft und Werkstofftechnik, 49[5] 2018, 689-695. https://doi.org/10.1002/mawe.201800010

[152] AlMangour, B., Yang, J.M., Materials and Design, 110, 2016, 914-924. https://doi.org/10.1016/j.matdes.2016.08.037

[153] Martina, F., Ding, J., Williams, S., Caballero, A., Pardal, G., Quintino, L., Additive Manufacturing, 25, 2019, 545-550. https://doi.org/10.1016/j.addma.2018.11.022

[154] Gonzalez-Gutierrez, J., Arbeiter, F., Schlauf, T., Kukla, C., Holzer, C., Materials Letters, 248, 2019, 165-168. https://doi.org/10.1016/j.matlet.2019.04.024

[155] Wang, Z., Wang, H., Liu, D., Chinese Journal of Lasers, 43[4] 2016, 0403001. https://doi.org/10.3788/CJL201643.0403001

[156] Yilmaz, O., Ugla, A.A., International Journal of Advanced Manufacturing Technology, 89[1-4] 2017, 13-25. https://doi.org/10.1007/s00170-016-9053-y

[157] Yang, X., Liu, J., Cui, X., Jin, G., Liu, Z., Chen, Y., Feng, X., Journal of Physics and Chemistry of Solids, 130, 2019, 210-216. https://doi.org/10.1016/j.jpcs.2019.03.001

[158] Mahbooba, Z., Thorsson, L., Unosson, M., Skoglund, P., West, H., Horn, T., Rock,

C., Vogli, E., Harrysson, O., Applied Materials Today, 11, 2018, 264-269.
https://doi.org/10.1016/j.apmt.2018.02.011

[159] Li, X., Tan, Y.H., Willy, H.J., Wang, P., Lu, W., Cagirici, M., Ong, C.Y.A., Herng, T.S., Wei, J., Ding, J., Materials and Design, 178, 2019, 107881.
https://doi.org/10.1016/j.matdes.2019.107881

[160] Zhao, X., Wei, Q.S., Gao, N., Zheng, E.L., Shi, Y.S., Yang, S.F., Journal of Materials Processing Technology, 270, 2019, 8-19.
https://doi.org/10.1016/j.jmatprotec.2019.01.028

[161] Zhang, X., Wang, K., Zhou, Q., Ding, J., Ganguly, S., Marzio, G., Yang, D., Xu, X., Dirisu, P., Williams, S.W., Materials Science and Engineering A, 762, 2019, 138097. https://doi.org/10.1016/j.msea.2019.138097

[162] Zhang, M., Chen, C., Qin, L., Yan, K., Cheng, G., Jing, H., Zou, T., Materials Science and Technology, 34[1] 2018, 69-78.
https://doi.org/10.1080/02670836.2017.1355584

[163] Bohlen, A., Freisse, H., Hunkel, M., Vollertsen, F., Procedia CIRP, 74, 2018, 192-195. https://doi.org/10.1016/j.procir.2018.08.092

[164] Zhang, X., Zhou, Q., Wang, K., Peng, Y., Ding, J., Kong, J., Williams, S., Materials and Design, 166, 2019, 107611.
https://doi.org/10.1016/j.matdes.2019.107611

[165] Duraisamy, R., Kumar, S.M., Kannan, A.R., Shanmugam, N.S., Sankaranarayanasamy, K., Proceedings of the Institution of Mechanical Engineers C, 2019, in press.

[166] Wu, M.W., Chen, J.K., Lin, B.H., Chiang, P.H., Materials and Design, 134, 2017, 163-170. https://doi.org/10.1016/j.matdes.2017.08.048

[167] Sander, J., Hufenbach, J., Giebeler, L., Wendrock, H., Kühn, U., Eckert, J., Materials and Design, 89, 2016, 335-341.
https://doi.org/10.1016/j.matdes.2015.09.148

[168] Xiong, J., Li, Y., Li, R., Yin, Z., Journal of Materials Processing Technology, 252, 2018, 128-136. https://doi.org/10.1016/j.jmatprotec.2017.09.020

[169] Hoefer, K., Nitsche, A., Abstoss, K.G., Ertugrul, G., Haelsig, A., Mayr, P., JOM, 71[4] 2019, 1554-1559. https://doi.org/10.1007/s11837-019-03356-4

[170] Liu, C., Zhang, M., Chen, C., Materials Science and Engineering A, 703, 2017, 359-371. https://doi.org/10.1016/j.msea.2017.07.031

[171] Palanivel, S., Sidhar, H., Mishra, R.S., JOM, 67[3] 2015, 616-621.
https://doi.org/10.1007/s11837-014-1271-x

[172] Palanivel, S., Nelaturu, P., Glass, B., Mishra, R.S., Materials and Design, 65, 2015,
934-952. https://doi.org/10.1016/j.matdes.2014.09.082

[173] Zhang, M., Chen, C., Liu, C., Wang, S., Metals, 8[8] 2018, 635.
https://doi.org/10.3390/met8080635

[174] Takagi, H., Sasahara, H., Abe, T., Sannomiya, H., Nishiyama, S., Ohta, S.,
Nakamura, K., Additive Manufacturing, 24, 2018, 498-507.
https://doi.org/10.1016/j.addma.2018.10.026

[175] Wolff, M., Mesterknecht, T., Bals, A., Ebel, T., Willumeit-Römer, R., Minerals,
Metals and Materials Series, 2019, 43-49. https://doi.org/10.1007/978-3-030-
05789-3_8

[176] Mahmood, S., Qureshi, A.J., Goh, K.L., Talamona, D., Rapid Prototyping Journal,
23[1] 2017, 122-128. https://doi.org/10.1108/RPJ-08-2015-0115

[177] Mahmood, S., Qureshi, A.J., Goh, K.L., Talamona, D., Rapid Prototyping Journal,
23[3] 2017, 524-533. https://doi.org/10.1108/RPJ-10-2015-0151

[178] Yap, C.Y., Tan, H.K., Du, Z., Chua, C.K., Dong, Z., Rapid Prototyping Journal,
23[4] 2017, 750-757. https://doi.org/10.1108/RPJ-01-2016-0006

[179] Luo, L., Gong, X., Russian Journal of Non-Ferrous Metals, 58[3] 2017, 269-275.
https://doi.org/10.3103/S1067821217030129

[180] Jia, W., Chen, S., Wei, M., Liang, J., Liu, C., Li, J., Powder Metallurgy, 62[1]
2019, 30-37. https://doi.org/10.1080/00325899.2018.1546921

[181] Li, Q., Lin, X., Liu, F., Liu, F., Huang, W., Materials Science and Engineering A,
700, 2017, 649-655. https://doi.org/10.1016/j.msea.2017.05.064

[182] Torgerson, T.B., Mantri, S.A., Banerjee, R., Scharf, T.W., Wear, 426-427, 2019,
942-951. https://doi.org/10.1016/j.wear.2018.12.046

[183] Wu, W., Jiang, J., Li, G., Fuh, J.Y.H., Jiang, H., Gou, P., Zhang, L., Liu, W., Zhao,
J., Journal of Non-Crystalline Solids, 506, 2019, 1-5.
https://doi.org/10.1016/j.jnoncrysol.2018.12.008

[184] Haberland, C., Elahinia, M., Walker, J.M., Meier, H., Frenzel, J., Smart Materials
and Structures, 23[10] 2014, 104002. https://doi.org/10.1088/0964-
1726/23/10/104002

[185] Khademzadeh, S., Zanini, F., Bariani, P.F., Carmignato, S., International Journal of

Advanced Manufacturing Technology, 96[9-12] 2018, 3729-3736.
https://doi.org/10.1007/s00170-018-1822-3

[186] Keshavarzkermani, A., Esmaeilizadeh, R., Ali, U., Enrique, P.D., Mahmoodkhani, Y., Zhou, N.Y., Bonakdar, A., Toyserkani, E., Materials Science and Engineering A, 762, 2019, 138081. https://doi.org/10.1016/j.msea.2019.138081

[187] Ramakrishnan, A., Dighe, A., Dinda, G., Materials Science and Technology, 2018, 125-132.

[188] Ramakrishnan, A., Dinda, G.P., Materials Science and Engineering A, 748, 2019, 347-356. https://doi.org/10.1016/j.msea.2019.01.101

[189] Li, S., Wei, Q., Prof., Shi, Y., Chua, C.K., Zhu, Z., Zhang, D., Journal of Materials Science and Technology, 31[9] 2015, 946-952. https://doi.org/10.1016/j.jmst.2014.09.020

[190] Lass, E.A., Stoudt, M.R., Williams, M.E., Katz, M.B., Levine, L.E., Phan, T.Q., Gnaeupel-Herold, T.H., Ng, D.S., Metallurgical and Materials Transactions A, 48[11] 2017, 5547-5558. https://doi.org/10.1007/s11661-017-4304-6

[191] Pleass, C., Jothi, S., Additive Manufacturing, 24, 2018, 419-431. https://doi.org/10.1016/j.addma.2018.09.023

[192] Cao, S., Gu, D., Journal of Materials Research, 30[23] 2015, 3616-3628. https://doi.org/10.1557/jmr.2015.358

[193] Hong, C., Gu, D., Dai, D., Alkhayat, M., Urban, W., Yuan, P., Cao, S., Gasser, A., Weisheit, A., Kelbassa, I., Zhong, M., Poprawe, R., Materials Science and Engineering A, 635, 2015, 118-128. https://doi.org/10.1016/j.msea.2015.03.043

[194] Gu, D., Cao, S., Lin, K., Journal of Manufacturing Science and Engineering, Transactions of the ASME, 139[4] 2017, 041014.

[195] Zhang, B., Bi, G., Chew, Y., Wang, P., Ma, G., Liu, Y., Moon, S.K., Applied Surface Science, 490, 2019, 522-534. https://doi.org/10.1016/j.apsusc.2019.06.008

[196] Gao, Y., Zhou, M., Applied Sciences, 8[12] 2018, 2439. https://doi.org/10.3390/app8122439

[197] List, F.A., Dehoff, R.R., Lowe, L.E., Sames, W.J., Materials Science and Engineering A, 615, 2014, 191-197. https://doi.org/10.1016/j.msea.2014.07.051

[198] Murr, L.E., Martinez, E., Gaytan, S.M., Ramirez, D.A., MacHado, B.I., Shindo, P.W., Martinez, J.L., Medina, F., Wooten, J., Ciscel, D., Ackelid, U., Wicker, R.B., Metallurgical and Materials Transactions A, 42[11] 2011, 3491-3508.

https://doi.org/10.1007/s11661-011-0748-2

[199] Gonzalez, J.A., Mireles, J., Stafford, S.W., Perez, M.A., Terrazas, C.A., Wicker, R.B., Journal of Materials Processing Technology, 264, 2019, 200-210. https://doi.org/10.1016/j.jmatprotec.2018.08.031

[200] Mostafaei, A., Stevens, E.L., Hughes, E.T., Biery, S.D., Hilla, C., Chmielus, M., Materials and Design, 108, 2016, 126-135. https://doi.org/10.1016/j.matdes.2016.06.067

[201] Enrique, P.D., Marzbanrad, E., Mahmoodkhani, Y., Jiao, Z., Toyserkani, E., Zhou, N.Y., Surface and Coatings Technology, 362, 2019, 141-149. https://doi.org/10.1016/j.surfcoat.2019.01.108

[202] Enrique, P.D., Mahmoodkhani, Y., Marzbanrad, E., Toyserkani, E., Zhou, N.Y., Materials Letters, 232, 2018, 179-182. https://doi.org/10.1016/j.matlet.2018.08.117

[203] Rivera, O.G., Allison, P.G., Jordon, J.B., Rodriguez, O.L., Brewer, L.N., McClelland, Z., Whittington, W.R., Francis, D., Su, J., Martens, R.L., Hardwick, N., Materials Science and Engineering A, 694, 2017, 1-9. https://doi.org/10.1016/j.msea.2017.03.105

[204] Yangfan, W., Xizhang, C., Chuanchu, S., Surface and Coatings Technology, 374, 2019, 116-123. https://doi.org/10.1016/j.surfcoat.2019.05.079

[205] Zhong, M., Yang, L., Liu, W., Huang, T., He, J., Proceedings of SPIE, 5629, 2005, 59-66.

[206] Liu, P., Sun, S.Y., Gong, J.H., Xu, S.B., Liu, X.N., Xuan, C., Wu, C.M., Lasers in Engineering, 43[1-3] 2019, 47-58.

[207] Ni, M., Chen, C., Wang, X., Wang, P., Li, R., Zhang, X., Zhou, K., Materials Science and Engineering A, 701, 2017, 344-351. https://doi.org/10.1016/j.msea.2017.06.098

[208] Jiang, R., Mostafaei, A., Pauza, J., Kantzos, C., Rollett, A.D., Materials Science and Engineering A, 755, 2019, 170-180. https://doi.org/10.1016/j.msea.2019.03.103

[209] Wang, Y., Shi, J., Lu, S., Wang, Y., ASME 11th International Manufacturing Science and Engineering Conference, 2016, 3.

[210] Wang, Y., Shi, J., Lu, S., Wang, Y., ASME 11th International Manufacturing Science and Engineering Conference, 2016, 1.

[211] Wang, Y., Shi, J., 13th International Manufacturing Science and Engineering Conference, 2018, 2.

[212] Witzel, J., Schopphoven, T., Gasser, A., Kelbassa, I., 30th International Congress on Applications of Lasers and Electro-Optics, 2011, 275-282.

[213] Kuo, Y.L., Horikawa, S., Kakehi, K., Materials and Design, 116, 2017, 411-418. https://doi.org/10.1016/j.matdes.2016.12.026

[214] Witkin, D.B., Hayes, R.W., McLouth, T.D., Bean, G.E., Metallurgical and Materials Transactions A, 50[8] 2019, 3458-3465. https://doi.org/10.1007/s11661-019-05298-7

[215] Onuike, B., Bandyopadhyay, A., Materials Letters, 252, 2019, 256-259. https://doi.org/10.1016/j.matlet.2019.05.114

[216] Baykasoglu, C., Akyildiz, O., Candemir, D., Yang, Q., To, A.C., Journal of Manufacturing Science and Engineering, Transactions of the ASME, 140[5] 2018, 0510031. https://doi.org/10.1115/1.4038894

[217] Kong, D., Dong, C., Ni, X., Zhang, L., Man, C., Yao, J., Ji, Y., Ying, Y., Xiao, K., Cheng, X., Li, X., Journal of Alloys and Compounds, 785, 2019, 826-837. https://doi.org/10.1016/j.jallcom.2019.01.263

[218] Onuike, B., Heer, B., Bandyopadhyay, A., Additive Manufacturing, 21, 2018, 133-140. https://doi.org/10.1016/j.addma.2018.02.007

[219] Jia, Q., Gu, D., Journal of Materials Research, 29[17] 2014, 1960-1969. https://doi.org/10.1557/jmr.2014.130

[220] Wang, Y., Shi, J., Deng, X., Lu, S., ASME International Mechanical Engineering Congress and Exposition, Proceedings, 2016, 2.

[221] Gu, D., Zhang, H., Dai, D., Xia, M., Hong, C., Poprawe, R., Composites B, 163, 2019, 585-597. https://doi.org/10.1016/j.compositesb.2018.12.146

[222] Al-Juboori, L.A., Niendorf, T., Brenne, F., Metallurgical and Materials Transactions B, 49[6] 2018, 2969-2974. https://doi.org/10.1007/s11663-018-1407-4

[223] Balachandramurthi, A.R., Moverare, J., Mahade, S., Pederson, R., Materials, 12[1] 2018, 68. https://doi.org/10.3390/ma12010068

[224] Sames, W.J., Unocic, K.A., Helmreich, G.W., Kirka, M.M., Medina, F., Dehoff, R.R., Babu, S.S., Additive Manufacturing, 13, 2017, 156-165. https://doi.org/10.1016/j.addma.2016.09.001

[225] Kirka, M.M., Lee, Y., Greeley, D.A., Okello, A., Goin, M.J., Pearce, M.T., Dehoff, R.R., JOM, 69[3] 2017, 523-531. https://doi.org/10.1007/s11837-017-2264-3

[226] Yamashita, Y., Murakami, T., Mihara, R., Okada, M., Murakami, Y., International Journal of Fatigue, 117, 2018, 485-495. https://doi.org/10.1016/j.ijfatigue.2018.08.002

[227] Yamashita, Y., Murakami, T., Mihara, R., Okada, M., Murakami, Y., Procedia Structural Integrity, 7, 2017, 11-18. https://doi.org/10.1016/j.prostr.2017.11.054

[228] Xu, X., Ding, J., Ganguly, S., Williams, S., Journal of Materials Processing Technology, 265, 2019, 201-209. https://doi.org/10.1016/j.jmatprotec.2018.10.023

[229] Lin, C.Y., Bor, H.Y., Wei, C.N., Liao, C.H., Materials Science Forum, 941, 2018, 2167-2172. https://doi.org/10.4028/www.scientific.net/MSF.941.2167

[230] Valdez, M., Kozuch, C., Faierson, E.J., Jasiuk, I., Journal of Alloys and Compounds, 725, 2017, 757-764. https://doi.org/10.1016/j.jallcom.2017.07.198

[231] Raghavan, N., Dehoff, R., Pannala, S., Simunovic, S., Kirka, M., Turner, J., Carlson, N., Babu, S.S., Acta Materialia, 112, 2016, 303-314. https://doi.org/10.1016/j.actamat.2016.03.063

[232] Xu, J., Lin, X., Guo, P., Yang, H., Xue, L., Huang, W., Journal of Alloys and Compounds, 2019, 461-475. https://doi.org/10.1016/j.jallcom.2018.11.386

[233] Asala, G., Khan, A.K., Andersson, J., Ojo, O.A., Metallurgical and Materials Transactions A, 48[9] 2017, 4211-4228. https://doi.org/10.1007/s11661-017-4162-2

[234] Marenych, O., Kostryzhev, A., Shen, C., Pan, Z., Li, H., van Duin, S., Metals, 9[1] 2019, 105. https://doi.org/10.3390/met9010105

[235] Martin, J.H., Ashby, D.S., Schaedler, T.A., Materials and Design, 120, 2017, 291-297. https://doi.org/10.1016/j.matdes.2017.02.023

[236] Gan, Z., Liu, H., Li, S., He, X., Yu, G., International Journal of Heat and Mass Transfer, 111, 2017, 709-722. https://doi.org/10.1016/j.ijheatmasstransfer.2017.04.055

[237] Guo, Y., Jia, L., Kong, B., Zhang, S., Zhang, F., Zhang, H., Applied Surface Science, 409, 2017, 367-374. https://doi.org/10.1016/j.apsusc.2017.02.221

[238] Sungail, C., Abid, A.D., Metal Powder Report, 2019, in press.

[239] Fang, X., Wei, Z., Du, J., Bingheng, L., He, P., Wang, B., Chen, J., Geng, R., Rapid Prototyping Journal, 23[5] 2017, 893-903. https://doi.org/10.1108/RPJ-03-

2016-0052

[240] Zhao, G., Wei, Z., Du, J., Liu, W., Wang, X., Yao, Y., Procedia Engineering, 157, 2016, 193-199. https://doi.org/10.1016/j.proeng.2016.08.356

[241] Wood, J.V., Materials Science and Technology, 4, 1988, 189-193 https://doi.org/10.1179/mst.1988.4.3.189

[242] Kim, H.K., Kim, H.G., Lee, B.S., Min, S.H., Ha, T.K., Jung, K.H., Lee, C.W., Park, H.K., Materials Transactions, 58[4] 2017, 592-595. https://doi.org/10.2320/matertrans.M2016361

[243] Traxel, K.D., Bandyopadhyay, A., Additive Manufacturing, 24, 2018, 353-363. https://doi.org/10.1016/j.addma.2018.10.005

[244] Gu, D., Wang, H., Zhang, G., Metallurgical and Materials Transactions A, 45[1] 2014, 464-476. https://doi.org/10.1007/s11661-013-1968-4

[245] Sheydaeian, E., Toyserkani, E., Composites B, 138, 2018, 140-148. https://doi.org/10.1016/j.compositesb.2017.11.035

[246] Hu, Y., Cong, W., Wang, X., Li, Y., Ning, F., Wang, H., Composites B, 133, 2018, 91-100. https://doi.org/10.1016/j.compositesb.2017.09.019

[247] Zhang, Y., Bandyopadhyay, A., Additive Manufacturing, 21, 2018, 104-111. https://doi.org/10.1016/j.addma.2018.03.001

[248] Mahamood, R.M., Akinlabi, E.T., IOP Conference Series - Materials Science and Engineering, 391[1] 2018, 012005. https://doi.org/10.1088/1757-899X/391/1/012005

[249] Avila, J.D., Bandyopadhyay, A., Journal of Materials Research, 34[7] 2019, 1279-1289. https://doi.org/10.1557/jmr.2019.11

[250] Xia, M., Liu, A., Hou, Z., Li, N., Chen, Z., Ding, H., Journal of Alloys and Compounds, 728, 2017, 436-444. https://doi.org/10.1016/j.jallcom.2017.09.033

[251] Farayibi, P.K., Abioye, T.E., Kennedy, A., Clare, A.T., Journal of Manufacturing Processes, 45, 2019, 429-437. https://doi.org/10.1016/j.jmapro.2019.07.029

[252] Polozov, I., Sufiiarov, V., Popovich, A., Borisov, E., Masaylo, D., Orlov, A., 27th International Conference on Metallurgy and Materials, 2018, 1677-1684.

[253] Ma, Y., Cuiuri, D., Li, H., Pan, Z., Shen, C., Materials Science and Engineering A, 657, 2016, 86-95. https://doi.org/10.1016/j.msea.2016.01.060

[254] Murr, L.E., Gaytan, S.M., Ceylan, A., Martinez, E., Martinez, J.L., Hernandez,

D.H., Machado, B.I., Ramirez, D.A., Medina, F., Collins, S., Wicker, R.B., Acta Materialia, 58[5] 2010, 1887-1894. https://doi.org/10.1016/j.actamat.2009.11.032

[255] Wu, Y., Zhang, S., Cheng, X., Wang, H., Journal of Alloys and Compounds, 2019, 799, 325-333. https://doi.org/10.1016/j.jallcom.2019.05.337

[256] Bakhshivash, S., Asgari, H., Russo, P., Dibia, C.F., Ansari, M., Gerlich, A.P., Toyserkani, E., International Journal of Advanced Manufacturing Technology, 103[9-12] 2019, 4399-4409. https://doi.org/10.1007/s00170-019-03847-3

[257] Kurzynowski, T., Madeja, M., Dziedzic, R., Kobiela, K., Scanning, 2019, 2903920.

[258] Kenel, C., Dawson, K., Barras, J., Hauser, C., Dasargyri, G., Bauer, T., Colella, A., Spierings, A.B., Tatlock, G.J., Leinenbach, C., Wegener, K., Intermetallics, 90, 2017, 63-73. https://doi.org/10.1016/j.intermet.2017.07.004

[259] Kenel, C., Lis, A., Dawson, K., Stiefel, M., Pecnik, C., Barras, J., Colella, A., Hauser, C., Tatlock, G.J., Leinenbach, C., Wegener, K., Intermetallics, 91, 2017, 169-180. https://doi.org/10.1016/j.intermet.2017.09.004

[260] Zhu, Y.Y., Chen, B., Tang, H.B., Cheng, X., Wang, H.M., Li, J., Transactions of Nonferrous Metals Society of China, 28[1] 2018, 36-46. https://doi.org/10.1016/S1003-6326(18)64636-9

[261] Chen, X., Zhang, J., Chen, X., Cheng, X., Huang, Z., Vacuum, 151, 2018, 116-121. https://doi.org/10.1016/j.vacuum.2018.02.011

[262] Bhardwaj, T., Shukla, M., Paul, C.P., Bindra, K.S., Journal of Alloys and Compounds, 787, 2019, 1238-1248. https://doi.org/10.1016/j.jallcom.2019.02.121

[263] Zhang, Y., Sahasrabudhe, H., Bandyopadhyay, A., Applied Surface Science, 346, 2015, 428-437. https://doi.org/10.1016/j.apsusc.2015.03.184

[264] Zhao, D., Han, C., Li, Y., Li, J., Zhou, K., Wei, Q., Liu, J., Shi, Y., Journal of Alloys and Compounds, 804, 2019, 288-298. https://doi.org/10.1016/j.jallcom.2019.06.307

[265] Baufeld, B., Biest, O.V.D., Gault, R., Materials and Design, 31[S1] 2010, S106-S111. https://doi.org/10.1016/j.matdes.2009.11.032

[266] Harun, W.S.W., Manam, N.S., Kamariah, M.S.I.N., Sharif, S., Zulkifly, A.H., Ahmad, I., Miura, H., Powder Technology, 331, 2018, 74-97. https://doi.org/10.1016/j.powtec.2018.03.010

[267] Larosa, M.A., Jardini, A.L., Zavaglia, C.A.D.C., Kharmandayan, P., Calderoni, D.R., Maciel Filho, R., Advances in Mechanical Engineering, 2014, 945819.

https://doi.org/10.1155/2014/945819

[268] Harlow, D.G., 21st ISSAT International Conference on Reliability and Quality in Design, 2015, 12-16.

[269] Dey, N.K., Liou, F.W., Nedic, C., 24th International SFF Symposium - An Additive Manufacturing Conference, SFF 2013, 853-858.

[270] Ladani, L., Razmi, J., Choudhury, S.F., Journal of Engineering Materials and Technology, Transactions of the ASME, 136[3] 2014, 031006.

[271] Li, N., Xiong, Y., Xiong, H., Shi, G., Blackburn, J., Liu, W., Qin, R., Materials Characterization, 148, 2019, 43-51. https://doi.org/10.1016/j.matchar.2018.11.032

[272] Das, M., Bhattacharya, K., Dittrick, S.A., Mandal, C., Balla, V.K., Sampath Kumar, T.S., Bandyopadhyay, A., Manna, I., Journal of the Mechanical Behavior of Biomedical Materials, 29, 2014, 259-271. https://doi.org/10.1016/j.jmbbm.2013.09.006

[273] Hattal, A., Chauveau, T., Djemai, M., Fouchet, J.J., Bacroix, B., Dirras, G., Materials and Design, 180, 2019, 107909. https://doi.org/10.1016/j.matdes.2019.107909

[274] Sahasrabudhe, H., Bandyopadhyay, A., Journal of the Mechanical Behavior of Biomedical Materials, 85, 2018, 1-11. https://doi.org/10.1016/j.jmbbm.2018.05.020

[275] Wysocki, B., Maj, P., Sitek, R., Buhagiar, J., Kurzydłowski, K.J., Świeszkowski, W., Applied Sciences, 7[7] 2017, 657. https://doi.org/10.3390/app7070657

[276] Khrapov, D., Surmeneva, M., Koptioug, A., Evsevleev, S., Lé Onard, F., Bruno, G., Surmenev, R., Journal of Physics - Conference Series, 1145[1] 2019, 012044. https://doi.org/10.1088/1742-6596/1145/1/012044

[277] Raghavan, A., Wei, H.L., Palmer, T.A., Debroy, T., Journal of Laser Applications, 25[5] 2013, 052006. https://doi.org/10.2351/1.4817788

[278] Ge, P., Zhang, Z., Tan, Z.J., Hu, C.P., Zhao, G.Z., Guo, X., International Journal of Thermal Sciences, 140, 2019, 329-343. https://doi.org/10.1016/j.ijthermalsci.2019.03.013

[279] Wei, L.C., Ehrlich, L.E., Powell-Palm, M.J., Montgomery, C., Beuth, J., Malen, J.A., Additive Manufacturing, 21, 2018, 201-208. https://doi.org/10.1016/j.addma.2018.02.002

[280] Mahamood, R.M., Akinlabi, E.T., Materials Today - Proceedings, 5[9] 2018,

18362-18367. https://doi.org/10.1016/j.matpr.2018.06.175

[281] Gockel, J., Klingbeil, N., Bontha, S., Metallurgical and Materials Transactions B, 47[2] 2016, 1400-1408. https://doi.org/10.1007/s11663-015-0547-z

[282] Mukherjee, T., Manvatkar, V., De, A., DebRoy, T., Journal of Applied Physics, 121[6] 2017, 064904. https://doi.org/10.1063/1.4976006

[283] Kurz, W., Fisher, D.J., Trivedi, R., International Materials Reviews, 64[6] 2019, 311-354. https://doi.org/10.1080/09506608.2018.1537090

[284] Lin, W.S., Starr, T.L., Harris, B.T., Zandinejad, A., Morton, D., International Journal of Oral and Maxillofacial Implants, 28[6] 2013, 1490-1495. https://doi.org/10.11607/jomi.3164

[285] Sartori, S., Bordin, A., Moro, L., Ghiotti, A., Bruschi, S., Procedia CIRP, 46, 2016, pp. 587-590. https://doi.org/10.1016/j.procir.2016.04.032

[286] Petrousek, P., Bidulska, J., Bidulsky, R., Kocisko, R., Fedorikova, A., Hudak, R., Rajtukova, V., Zivcak, J., MM Science Journal, 2017, 1752-1755. https://doi.org/10.17973/MMSJ.2017_02_2016190

[287] Longhitano, G.A., Arenas, M.A., Conde, A., Larosa, M.A., Jardini, A.L., Zavaglia, C.A.D.C., Damborenea, J.J., Journal of Alloys and Compounds, 765, 2018, 961-968. https://doi.org/10.1016/j.jallcom.2018.06.319

[288] Lin, J.J., Lv, Y.H., Liu, Y.X., Xu, B.S., Sun, Z., Li, Z.G., Wu, Y.X., Materials and Design, 102, 2016, 30-40. https://doi.org/10.1016/j.matdes.2016.04.018

[289] Spranger, F., Graf, B., Schuch, M., Hilgenberg, K., Rethmeier, M., Journal of Laser Applications, 30[2] 2018, 022001. https://doi.org/10.2351/1.4997852

[290] Peyre, P., Dal, M., Pouzet, S., Castelnau, O., Journal of Laser Applications, 29[2] 2017, 022304. https://doi.org/10.2351/1.4983251

[291] Longhitano, G.A., Larosa, M.A., Jardini, A.L., Zavaglia, C.A.D.C., Ierardi, M.C.F., Journal of Materials Processing Technology, 252, 2018, 202-210. https://doi.org/10.1016/j.jmatprotec.2017.09.022

[292] Wauthle, R., Ahmadi, S.M., Amin Yavari, S., Mulier, M., Zadpoor, A.A., Weinans, H., Van Humbeeck, J., Kruth, J.P., Schrooten, J., Materials Science and Engineering C, 54, 2015, 94-100. https://doi.org/10.1016/j.msec.2015.05.001

[293] Tuomi, J.T., Björkstrand, R.V., Pernu, M.L., Salmi, M.V.J., Huotilainen, E.I., Wolff, J.E.H., Vallittu, P.K., Mäkitie, A.A., Journal of Materials Science - Materials in Medicine, 28[3] 2017, 53. https://doi.org/10.1007/s10856-017-5863-1

[294] Nassar, A.R., Reutzel, E.W., Metallurgical and Materials Transactions A, 46[6] 2015, 2781-2789. https://doi.org/10.1007/s11661-015-2838-z

[295] Kakiuchi, T., Kawaguchi, R., Nakajima, M., Hojo, M., Fujimoto, K., Uematsu, Y., International Journal of Fatigue, 126, 2019, 55-61. https://doi.org/10.1016/j.ijfatigue.2019.04.025

[296] Lyczkowska, E., Szymczyk, P., Dybała, B., Chlebus, E., Archives of Civil and Mechanical Engineering, 14[4] 2014, 586-594. https://doi.org/10.1016/j.acme.2014.03.001

[297] Meier, C., Weissbach, R., Weinberg, J., Wall, W.A., Hart, A.J., Journal of Materials Processing Technology, 266, 2019, 484-501. https://doi.org/10.1016/j.jmatprotec.2018.10.037

[298] Meier, C., Weissbach, R., Weinberg, J., Wall, W.A., John Hart, A., Powder Technology, 343, 2019, 855-866. https://doi.org/10.1016/j.powtec.2018.11.072

[299] Desai, P.S., Mehta, A., Dougherty, P.S.M., Higgs, F.C., Powder Technology, 342, 2019, 441-456. https://doi.org/10.1016/j.powtec.2018.09.047

[300] Fiaz, H.S., Settle, C.R., Hoshino, K., Sensors and Actuators, A, 249, 2016, 284-293. https://doi.org/10.1016/j.sna.2016.08.029

[301] Gong, X., Chou, K., JOM, 67[5] 2015, 1176-1182. https://doi.org/10.1007/s11837-015-1352-5

[302] Tang, H.P., Qian, M., Liu, N., Zhang, X.Z., Yang, G.Y., Wang, J., JOM, 67[3] 2015, 555-563. https://doi.org/10.1007/s11837-015-1300-4

[303] Wei, C., Ma, X., Yang, X., Zhou, M., Wang, C., Zheng, Y., Zhang, W., Li, Z., Materials Letters, 221, 2018, 111-114. https://doi.org/10.1016/j.matlet.2018.03.124

[304] Sun, Y., Aindow, M., Hebert, R.J., Materials at High Temperatures, 35[1-3] 2018, 217-224. https://doi.org/10.1080/09603409.2017.1389133

[305] Tshabalala, L., Mathe, N., Chikwanda, H., Key Engineering Materials, 770, 2018, 3-8. https://doi.org/10.4028/www.scientific.net/KEM.770.3

[306] Price, S., Cheng, B., Lydon, J., Cooper, K., Chou, K., Journal of Manufacturing Science and Engineering, Transactions of the ASME, 136[6] 2014, 061019. https://doi.org/10.1115/1.4028485

[307] Madigan, R.B., Riley, S.F., Cola, M.J., Dave, V.R., Talkington, J.E., ASM Proceedings of the International Conference: Trends in Welding Research, 2013,

963-969.

[308] Vastola, G., Zhang, G., Pei, Q.X., Zhang, Y.W., JOM, 68[5] 2016, 1370-1375. https://doi.org/10.1007/s11837-016-1890-5

[309] Klocke, F., Arntz, K., Klingbeil, N., Schulz, M., Proceedings of SPIE, 2017, 100950U.

[310] Hoefer, K., Mayr, P., Materials Science Forum, 941, 2018, 2137-2141. https://doi.org/10.4028/www.scientific.net/MSF.941.2137

[311] Yang, G., Wang, B., Qin, L., Li, C., Wang, C., Chinese Journal of Rare Metals, 42[9] 2018, 903-908.

[312] Bermingham, M.J., Nicastro, L., Kent, D., Chen, Y., Dargusch, M.S., Journal of Alloys and Compounds, 2018, 753, 247-255. https://doi.org/10.1016/j.jallcom.2018.04.158

[313] Zhang, F., Chen, W., Tian, M., Rare Metal Materials and Engineering, 47[6] 2018, 1890-1895.

[314] Wu, B., Pan, Z., Ding, D., Cuiuri, D., Li, H., Additive Manufacturing, 23, 2018, 151-160. https://doi.org/10.1016/j.addma.2018.08.004

[315] Hu, R., Chen, X., Yang, G., Gong, S., Pang, S., International Journal of Heat and Mass Transfer, 126, 2018, 877-887. https://doi.org/10.1016/j.ijheatmasstransfer.2018.06.033

[316] Shi, X., Ma, S., Liu, C., Wu, Q., Lu, J., Liu, Y., Shi, W., Materials Science and Engineering A, 684, 2017, 196-204. https://doi.org/10.1016/j.msea.2016.12.065

[317] Mahbooba, Z., West, H., Harrysson, O., Wojcieszynski, A., Dehoff, R., Nandwana, P., Horn, T., JOM, 69[3] 2017, 472-478. https://doi.org/10.1007/s11837-016-2210-9

[318] Gorsse, S., Hutchinson, C., Gouné, M., Banerjee, R., Science and Technology of Advanced Materials, 18[1] 2017, 584-610. https://doi.org/10.1080/14686996.2017.1361305

[319] Gou, J., Shen, J., Hu, S., Tian, Y., Liang, Y., Journal of Manufacturing Processes, 42, 2019, 41-50. https://doi.org/10.1016/j.jmapro.2019.04.012

[320] Chen, C., Xie, Y., Yan, X., Yin, S., Fukanuma, H., Huang, R., Zhao, R., Wang, J., Ren, Z., Liu, M., Liao, H., Additive Manufacturing, 27, 2019, 595-605. https://doi.org/10.1016/j.addma.2019.03.028

[321] Petrovskiy, P., Sova, A., Doubenskaia, M., Smurov, I., International Journal of

Advanced Manufacturing Technology, 102[1-4] 2019, 819-827.
https://doi.org/10.1007/s00170-018-03233-5

[322] Chern, A.H., Nandwana, P., Yuan, T., Kirka, M.M., Dehoff, R.R., Liaw, P.K.,
Duty, C.E., International Journal of Fatigue, 119, 2019, 173-184.
https://doi.org/10.1016/j.ijfatigue.2018.09.022

[323] Morton, P.A., Mireles, J., Mendoza, H., Cordero, P.M., Benedict, M., Wicker,
R.B., Journal of Mechanical Design, Transactions of the ASME, 137[11] 2015,
114501. https://doi.org/10.1115/1.4031057

[324] Gibson, T., Tandon, G.P., Hick, A., Middendorf, J., Laycock, B., Simon, G.,
Conference Proceedings of the Society for Experimental Mechanics, 2, 2016, 91-
99. https://doi.org/10.1007/978-3-319-22443-5_11

[325] Faizan-Ur-Rab, M., Zahiri, S.H., King, P.C., Busch, C., Masood, S.H., Jahedi, M.,
Nagarajah, R., Gulizia, S., Journal of Thermal Spray Technology, 26[8] 2017,
1874-1887. https://doi.org/10.1007/s11666-017-0628-4

[326] Faizan-Ur-Rab, M., Zahiri, S.H., Masood, S.H., Jahedi, M., Nagarajah, R., TMS
Annual Meeting, 2016, 213-220. https://doi.org/10.1007/978-3-319-65133-0_26

[327] Wang, Z., Liu, P., Xiao, Y., Cui, X., Hu, Z., Chen, L., Journal of Manufacturing
Science and Engineering, Transactions of the ASME, 141[8] 2019, 081004.
https://doi.org/10.1115/1.4043798

[328] Koester, L.W., Taheri, H., Bond, L.J., Faierson, E.J., AIP Conference Proceedings,
2102, 2019, 020005.

[329] Abramovich, H., Broitman, N., Shirizly, A., 31st Congress of the International
Council of the Aeronautical Sciences, ICAS, 2018, 1-9.

[330] Lewandowski, J.J., Seifi, M., Annual Review of Materials Research, 46, 2016, 151-
186. https://doi.org/10.1146/annurev-matsci-070115-032024

[331] Fuerst, J., Medlin, D., Carter, M., Sears, J., Vander Voort, G., JOM, 67[4] 2015,
775-780. https://doi.org/10.1007/s11837-015-1345-4

[332] Popovich, V.A., Borisov, E.V., Heurtebise, V., Riemslag, T., Popovich, A.A.,
Sufiiarov, V.S., Minerals, Metals and Materials Series, F12, 2018, 85-97.
https://doi.org/10.1007/978-3-319-72526-0_9

[333] Palanivel, S., Dutt, A.K., Faierson, E.J., Mishra, R.S., Materials Science and
Engineering A, 654, 2016, 39-52. https://doi.org/10.1016/j.msea.2015.12.021

[334] Haque, M.S., Arrieta, E., Mireles, J., Carrasco, C., Stewart, C.M., Wicker, R.B.,

ASME International Mechanical Engineering Congress and Exposition, 2016, 9.

[335] Hrabe, N., Quinn, T., Materials Science and Engineering A, 573, 2013, 271-277. https://doi.org/10.1016/j.msea.2013.02.065

[336] Hrabe, N., Quinn, T., Materials Science and Engineering A, 573, 2013, 264-270. https://doi.org/10.1016/j.msea.2013.02.064

[337] Ge, W., Lin, F., Guo, C., Materials and Manufacturing Processes, 33[15] 2018, 1708-1713. https://doi.org/10.1080/10426914.2015.1048463

[338] Hutasoit, N., Masood, S.H., Pogula, K.S., Shuva, M.A.H., Rhamdhani, M.A., IOP Conference Series - Materials Science and Engineering, 377[1] 2018, 012138. https://doi.org/10.1088/1757-899X/377/1/012138

[339] Lin, J., Lv, Y., Liu, Y., Sun, Z., Wang, K., Li, Z., Wu, Y., Xu, B., Journal of the Mechanical Behavior of Biomedical Materials, 69, 2017, 19-29. https://doi.org/10.1016/j.jmbbm.2016.12.015

[340] Feng, X., Gu, H., Zhou, S., Lei, J., Chinese Journal of Lasers, 46[3] 2019, 0302003. https://doi.org/10.3788/CJL201946.0302003

[341] Rubino, F., Scherillo, F., Franchitti, S., Squillace, A., Astarita, A., Carlone, P., Journal of Manufacturing Processes, 37, 2019, 392-401. https://doi.org/10.1016/j.jmapro.2018.12.015

[342] Liu, Q., Wang, Y., Zheng, H., Tang, K., Ding, L., Li, H., Gong, S., Materials Science and Engineering A, 660, 2016, 24-33. https://doi.org/10.1016/j.msea.2016.02.069

[343] Tan, C., Zhou, K., Ma, W., Attard, B., Zhang, P., Kuang, T., Science and Technology of Advanced Materials, 19[1] 2018, 370-380. https://doi.org/10.1080/14686996.2018.1455154

[344] Guo, M., Gu, D., Xi, L., Zhang, H., Zhang, J., Yang, J., Wang, R., International Journal of Refractory Metals and Hard Materials, 84, 2019, 105025. https://doi.org/10.1016/j.ijrmhm.2019.105025

[345] Zhang, G., Gu, D., Rare Metal Materials and Engineering, 44[4] 2015, 1017-1023.

[346] Shuai, C., Cheng, Y., Yang, Y., Peng, S., Yang, W., Qi, F., Journal of Alloys and Compounds, 2019, 798, 606-615. https://doi.org/10.1016/j.jallcom.2019.05.278

[347] Wen, P., Voshage, M., Jauer, L., Chen, Y., Qin, Y., Poprawe, R., Schleifenbaum, J.H., Materials and Design, 155, 2018, 36-45. https://doi.org/10.1016/j.matdes.2018.05.057

[348] Wen, P., Qin, Y., Chen, Y., Voshage, M., Jauer, L., Poprawe, R., Schleifenbaum, J.H., Journal of Materials Science and Technology, 35[2] 2019, 368-376. https://doi.org/10.1016/j.jmst.2018.09.065

[349] Qin, Y., Wen, P., Voshage, M., Chen, Y., Schückler, P.G., Jauer, L., Xia, D., Guo, H., Zheng, Y., Schleifenbaum, J.H., Materials and Design, 181, 2019, 107937. https://doi.org/10.1016/j.matdes.2019.107937

[350] Sahasrabudhe, H., Bandyopadhyay, A., JOM, 68[3] 2016, 822-830. https://doi.org/10.1007/s11837-015-1777-x

Additive Manufacturing of Metals
Materials Research Foundations **67** (2020)

Materials Research Forum LLC
https://doi.org/10.21741/9781644900635

# Keyword Index

## About the author

**Dr Fisher** has wide knowledge and experience of the fields of engineering, metallurgy and solid-state physics, beginning with work at Rolls-Royce Aero Engines on turbine-blade research, related to the Concord supersonic passenger-aircraft project, which led to a BSc degree (1971) from the University of Wales. This was followed by theoretical and experimental work on the directional solidification of eutectic alloys having the ultimate aim of developing composite turbine blades. This work led to a doctoral degree (1978) from the Swiss Federal Institute of Technology (Lausanne). He then acted for many years as an editor of various academic journals, in particular *Defect and Diffusion Forum*. In recent years he has specialised in writing monographs which introduce readers to the most rapidly developing ideas in the fields of engineering, metallurgy and solid-state physics. His latest paper will appear shortly in *International Materials Reviews*, and he is co-author of the widely-cited student textbook, *Fundamentals of Solidification*.